Lyrical Earth Science

Text and Lyrics by Dorry and Doug Eldon

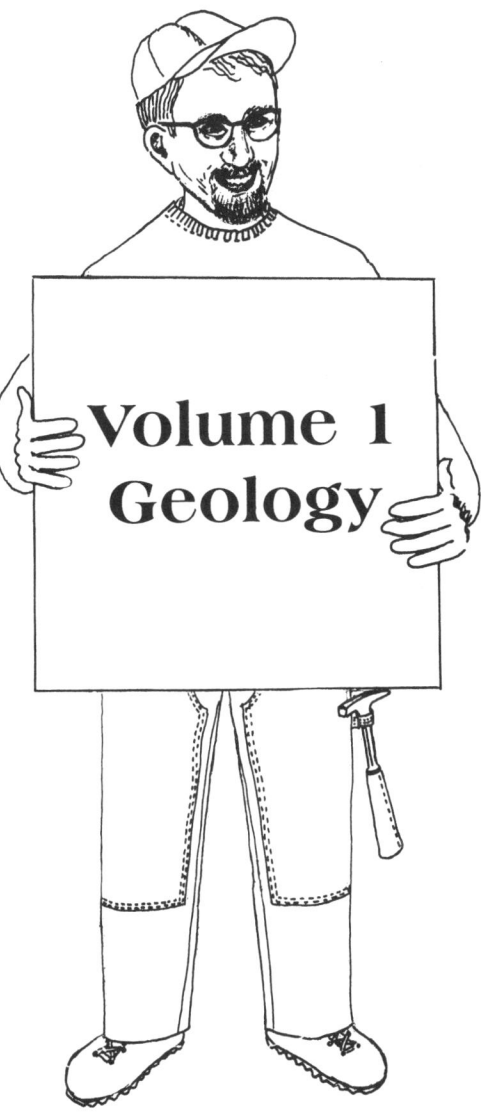

Volume 1
Geology

Illustrations by Sally Raskauskas

with assistance from Peter Wampler

Published by: Lyrical Learning
                8008 Cardwell Hill Dr.
                Corvallis, OR 97330
                541-754-3579 or 800-761-0906

> These book and CD sets can be used by students of many ages. Younger elementary students can benefit by becoming familiar with scientific terms through casual listening, yet may not fully understand the information until later. These songs, texts, and workbooks were written with middle school students in mind, as an introduction to the language of science.
>
> Workbooks are also available for each volume. Ideally, these resources should be in addition to hands-on activities where observations and applications can be made. In this way the knowledge learned through singing and reading can also become known through experience.
>
> Our website www.lyricallearning.com has reviews and awards; FAQs; an article about the theory and research behind using lyrics and music for learning; information about our three Lyrical Life Science volumes, as well as Lyrical Life Science Volume 1 revisions and updates; a list of our distributors and ordering information; and where to find our products as digital downloads.

Text and lyrics: Dorry and Doug Eldon   Layout and design: Dorry Eldon
Cover art and design: Bekki Levien   Illustrations: Sally Raskaukas
Sheet music layout: Noah Dietterich   Scientific advisor: Peter Wampler

Illustrations were mostly redrawn from the U.S. Geological Survey website and publications.

Illustrations on page 86 by Eric Altendorf, from Lyrical Life Science Volume 2: Mammals.

Lyrical Earth Science Volume 1: Geology text, CD and workbook set: ISBN 0-9-741635-1-1

Copyright Lyrical Learning
All rights reserved. The lyric sheets in the appendix may be reproduced by the purchaser for classroom or family use.

Printed in the USA
1st printing 2003
Reprinted 2006 and 2013

TABLE OF CONTENTS

Introduction		4
Lanforms and Geologic Features		5

Building Up the Crust—and Moving It Around

Chapter 1   **An Introduction to Geology**   6
     J Harlan Bretz, Geologic Laws

Chapter 2   **Plate Tectonics**   12
     Alfred Wegener, Parts of the Earth, Tectonic Plates, Plate Boundaries

Chapter 3   **Folding, Faulting, and Intrusion**   18
     Direction of Force, Folding, Faulting, Intrusion

Chapter 4   **Earthquakes**   24
     Location, Seismic Waves, Other Seismic Events

Chapter 5   **Volcanoes**   28
     Classification, Mafic Lava, Felsic Lava, Intermediate Lava, Other Forms of Volcanic Activity

What the Crust is made of—Rocks and Minerals

Chapter 6   **Minerals**   36
     A Little Chemistry, Characteristics, Classifying Mineral Compounds, Mineral Groupings, Industrial Uses, Types of Rock, The Rock Cycle

Chapter 7   **Igneous Rock**   42
     Types of Igneous Rock, Classification, Chemical Composition

Chapter 8   **Sedimentary Rock**   46
     Bedding, Kinds of Sedimentary Rock

Chapter 9   **Metamorphic Rock**   50
     Types of Metamorphism, Classification

Wearing Down the Crust

Chapter 10   **Weathering of Rocks**   54
     Conditions Affecting Rate of Weathering, Physical Weathering, Chemical Weathering

Chapter 11   **Hydrology and Erosion**   58
     The Water Cycle, Runoff and Erosion, Rate of Runoff, Deposition

Chapter 12   **Waterways and Erosion**   62
     Processes of Erosion, Rivers, Floodplain Characteristics

Chapter 13   **Groundwater**   66
     Porosity and Permeability, Infiltration, Ground Water Erosion

Chapter 14   **Mass Movement**   70
     Slope Conditions, Classification, Creep. Flows, Rockfalls and Rockslides, Subsidence

Chapter 15   **Glaciers**   74
     Kinds of Glaciers, Growth and Shrinkage, Movement, Erosion, Sediment and Meltwater Landforms from Till, Landforms from Outwash, A Word about Permafrost

Chapter 16   **Wind Erosion**   80
     Factors Affecting Rate of Wind Erosion, Causes of High Wind Erosion, Landforms Created by Wind Erosion

Chapter 17   **Soil**   84
     Weathering of Rock into Soil, Organic Matter and Organisms, Soil Profiles, Soil Groups

More Information

Chapter 18   **Topographic Maps**   88

Appendix   93
     Lyric Sheets, Notes. Bibliography. Index

INTRODUCTION

Welcome to *Lyrical Earth Science Volume 1: Geology*—a collection of 18 songs and chapters introducing you to the study of geology. We hope you not only enjoy the subject but grasp these basic geological fundamentals to give you a new appreciation for the world around you.

Geology is a subject of depth, breadth and complexity! Thankfully we have a good friend who could explain, and explain again, the terms and concepts. Peter Wampler, within a few quarters from completing his PhD. in geology, spent hours talking with us about the fundamentals and the details of a subject he loves. He gave additional time in reading and editing this manuscript between his own research, classes, presentations—and rock-hounding with his family. His input was invaluable for accuracy and clarity.

In addition to Peter's help, Bobby Horton, a talented and joyous musician, joins us once again. He pulls out all the stops to create fun and sometimes goofy songs to liven up your study of geology and help you through technical and sometimes complex geological information. He sets aside his traditional period arrangements used for his Civil War albums and public television historical documentaries. Now "anything goes" for the sake of making learning fun.

You may wonder where we get these old melodies. Well, *Blow Ye Winds in the Morning*, *My Grandfather's Clock*, and *The Boll Weevil* came from early elementary school days when the visiting music teacher tried her best to acquaint us with our musical heritage. Civil War tunes such as *Bonnie Blue Flag*, *Battle Cry of Freedom*, *Invalid Corps*, and *Just Before the Battle, Mother*, came from early and fond family memories listening to record albums produced for the centennial anniversary. *The Minstrel Boy* and *Wearing of the Green* are more lovely Irish melodies gathered over time; and *Columbia, the Gem of the Ocean*, was noted so often in our historical readings, that we knew we really needed it. *Anchors Away* was subconsciously, or otherwise, learned from a father who had been in the Navy. And on it goes... forming a collection of well loved, and once well known tunes of mostly bygone eras. We hope as you study geology, you will also come away with an appreciation for the breadth and emotion of music from long ago.

LANDFORMS AND GEOLOGIC FEATURES

1. river
2. river mouth
3. delta
4. meander
5. oxbow lake
6. confluence
7. tributary
8. floodplain
9. waterfalls
10. rapids
11. river gorge
12. headwater
13. alluvial fan
14. cliffs
15. mesa
16. butte
17. volcano
18. lava flow
19. crater
20. mountains
21. horn
22. glacier
23. hills
24. sand dunes
25. sedimentary beds
26. folded beds
27. syncline
28. anticline
29. normal fault
30. hanging wall
31. foot wall
32. cave
33. fault-block mountains
34. valley

AN INTRODUCTION TO GEOLOGY

(to the tune of "Battle Cry of Freedom")
This tune was written during the Civil War and quickly became one of the most favorite marching songs of Union soldiers.

So with history it is, both of humans and the earth
Searching the past for explanations
In geology we know all the effort makes it worth
Our searching the past for explanations
   Chorus: Landforms and fossils, from mountains to old bones
   Are geologic puzzles and riddles in the stone
   So we deduce and we infer what we think had to occur
   From searching the past for explanations

The geologic processes that we see today
Searching the rock for explanations
Some think to have occurred so long ago in the same way
Searching the rock for explanations
   Chorus: "Uniformitarianism" things change the same
   But "catastrophism" says major events came
   An upheaval: an earthquake, or eruption, or a flood,
   We're still searching the rock for explanations

Deposits of new sediment are on top of older layers
Searching the rock for explanations
So "relative age" means just the age, but not in years
Searching the rock for explanations
   Chorus: "Superposition" means older rock below
   Comparison by age but it doesn't help us know
   The absolute age; so they estimate the date
   From searching the rock for explanations

AN INTRODUCTION TO GEOLOGY

Life Science
(Biology)
   zoology
   botany
   microbiology
   bacteriology
Earth Science
   geology
   meteorology
   oceanography
   astronomy
Physical Science
   chemistry
   physics

Most of us are curious about the world around us—we ask questions about who we are, where we came from, and why things are the way they are. People throughout time have been fascinated by these questions, and scientists spend much time, money, and energy seeking answers, searching for explanations, and trying to develop an understanding of our world and how it works. The word "science," in fact, comes from the word meaning "to know," and scientists desire to know the answers to their questions.

Earth science is the study of nonliving things around us, as opposed to biology, the study of living things. Earth science includes **geology**, the study of the earth itself; **meteorology**, or atmospheric science, the study of weather and climate; **oceanography**, which includes both the living and nonliving things in the ocean; and **astronomy**, the study of what is beyond the earth in outer space.

Geology is science that deals with the history of the earth. The evidence we interpret to help us understand the story of the earth is written in the landforms we see and in the processes we observe. To gain the understanding we desire, including explanations of why the earth around us is the way it is, it's important to have basic knowledge as a foundation to build upon. From facts about minerals, rocks, and landforms, and the processes that build them up and wear them away, we are able to piece together explanations, called **theories**, about the past.

Discoveries are made and theories change as scientists use the scientific method, which includes the following steps:
  1-ask a question
  2-gather information
  3-make a hypothesis
  4-experiment
  5-analyze data
  6-state a conclusion

A theory is not a fact; it is an interpretation of facts. New facts might support a theory, or they might cause the theory to be questioned and eventually changed. To **deduce** means to take a theory or general principle and build meaning out of facts; to **infer** means to use known facts to explain what might have happened. Both involve using the steps of the **scientific method** with careful observation, logical reasoning, and even mental flexibility. In geology, the scientific method is circular in that once a hypothesis is tested, it may raise new questions and the process begins again.

Change in accepted theories often comes slowly with much research, new discoveries, and solid evidence.

Geologists study the earth using a four-step version of the scientific method.
  1-observe
  2-question-
    make a hypothesis
  3-predict
  4-test
As a hypothesis is tested, questions may be raised that change the hypothesis, creating a circular process as steps 2–4 are repeated.

Take, for example, the theory of **uniformitarianism**, which states that geological changes are uniform, consistent, and gradual over long periods of time. —A theory that began to be accepted in the mid 1700s but was radically changed by the mid 1900s.

J HARLAN BRETZ

In the early 1920s, a high school biology teacher named J Harlan Bretz was following his growing interest in geology by taking trips around central Washington state. Bretz had some basic knowledge and understanding about certain geologic processes: he knew how rivers flow and flood, and how flowing water washes away earth material and deposits it elsewhere. He was trying to make sense of what he saw there, but the evidence didn't fit the current accepted theory of uniformitarianism. He found six geologic features he could not interpret or explain unless he theorized a **catastrophe**, a violent, sudden event like a flood, earthquake, or volcanic eruption. But a catastrophe the size Bretz envisioned did not fit with the current theory of uniformitarianism.

J Harlan Bretz 1882–1991

Why no period after the "J"? Nobody knows— and nobody knows what it stands for— not even his geology students at the University of Chicago, where the colorful and crusty J Harlan Bretz taught geology during his career.

First, Bretz knew that when water flows over a cliff, the force of the falling water gouges out a **plunge pool** at the base of the **waterfall**. The plunge pools Bretz found, however, were HUGE lakes at the base of now dry cliffs.

Second, Bretz knew that rivers gouged **potholes** in the hard rock they flowed over, but the potholes he found were enormous. And **third**, he understood how rivers sometimes become a network of **braided channels** instead of a nice, neat river course. But the braided channels he found were immense—they covered hundreds of square miles. So he called them "braided scablands."

Waterfalls carve plunge pools.

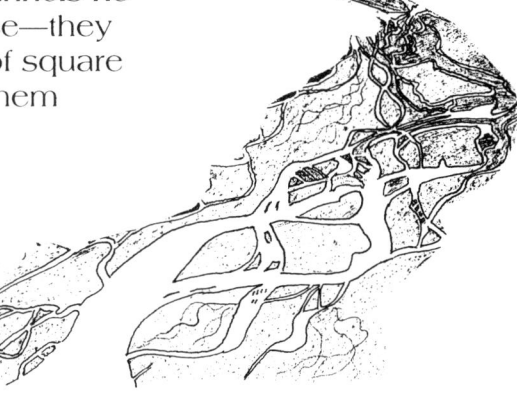
Braided channels in a riverbed.

An Introduction to Geology

Fourth, he was familiar with **gravel bars**, mounds of gravel that form in a river bed as the water drops, or slows down. But the gravel bars he found were hills the size of battleships!

Bretz saw gravel bars the size of battleships!

Fifth, he knew that when boulders get pushed downstream by rivers, they become rounded and worn down by **abrasion**. They are left where the water slows down and loses the force to push them any farther. But he found, instead, boulders with sharp, angled edges left on hillsides where there was no other evidence of flowing water—boulders the size of cars and houses! These rocks he called "**erratics**" because they were different from any other rocks found in the area. In fact, some erratics were the same as rocks only found hundreds of miles away! He knew glaciers (slow-moving masses of ice) carried and deposited erratics, but there were no other signs of glaciers!

And **sixth**, he found creeks and streams flowing into the Columbia River that were hundreds of feet above the level of the great river, a phenomenon which he knew was not normal. These "hanging valleys" were not the result of glaciers; in fact they did not fit anything he had ever seen.

Bretz tried to explain these six geologic features using the knowledge geologists had at that time. But he just couldn't make sense out of it all; he could not interpret or explain the evidence by the slow uniform process envisioned by uniformitarianism. Geologists are usually able to come up with some idea of what happened to create landforms, but what Bretz saw was so incredibly huge that his brain had to come up with a new explanation!

Bretz saw boulders the size of houses and cars!

Bretz incorporated these six geologic features into a theory involving a **cataclysm**, or an overwhelming and incredibly destructive flood. He estimated the flood to be ten times the amount of water flowing in all the rivers of the world today, traveling 50 to 60 miles an hour (96 kmh), at a depth of 200 to 500 feet (60–152 m)! The story he proposed involved a flood so massive, so powerful, and so unlike anything geologists had ever seen that they simply refused to listen to him.—Who was this biologist trying to change their theory of uniformitarianism? His ideas were rejected until more evidence was found. A source of water was discovered years later that could create

Bretz theorized a 200-foot (60m) wall of water traveling at 60 miles per hour! (96 kmh)

such a flood, or floods—ancient Lake Missoula, once near present day Missoula, Montana.

GEOLOGIC LAWS

With evidence for the **Missoula**, or **Bretz Floods**, the theory of uniformitarianism was modified, and credibility and acceptance was given to the much older theory of catastrophism, or change from catastrophic earth events. The following geologic laws reflect centuries of work by geologists, and also include the relatively recent findings of J Harlan Bretz.

> A **hypothesis** may become a **theory** which may become a **law**.

1-**Uniformitarianism** states that under normal conditions, the way the earth changes now is the way it changed in the past—"the present is the key to the past." Change is usually uniform, consistent, gradual, and takes place over long periods of time.

2-**Catastrophism** takes into account major events, or catastrophes, such as massive floods, major volcanic eruptions, and devastating earthquakes.

3-**Original horizontality** explains that **sediments** (rock particles of various sizes) are first laid down, or **deposited**, in horizontal layers.

4-**Superposition** states that layers of older rocks are under those of newer rocks—if left undisturbed. (Pressure may later change this order, in a process called **deformation**.) Rocks get broken down into smaller particles and are washed away by water. They are deposited in a lower place than where they started, which causes the newer rock to rest on top.

> Pronunciations of some geologic law terms:
> **Uniformitarianism**—
>   U-ne-form-i-TAIR-e-un-iz-um
> **Catastrophism**—
>   Ka-TA-stro-fiz-um
> **Original Horizontality**—
>   O-RI-gi-nol
>   Hore-ez-on-TAL-i-tee
> **Superposition**—
>   SOO-per-po-zish-un
> **Deformation**
>   De-for-MA-shun

Superposition seems to occur naturally even in a pile of newspapers! The oldest will be on the bottom of the pile—unless something moves it.

Because older rock is usually on the bottom (unless there have been catastrophes or deformations), geologists can determine which rock is older, or its **relative age**—its age in relation to other rocks. Relative age is deduced by studying geologic features and applying geologic laws and principles. For example, young mountains are usually steep and angular, while old mountains are worn down and rounded by water and wind.

Just as you may not know the exact ages of people, you can probably deduce which ones are older by observing physical characteristics of aging.

Just as we do not know the **absolute age**, the age in years, of landforms and features, we can observe and deduce their **relative age** by applying geologic laws.

An Introduction to Geology

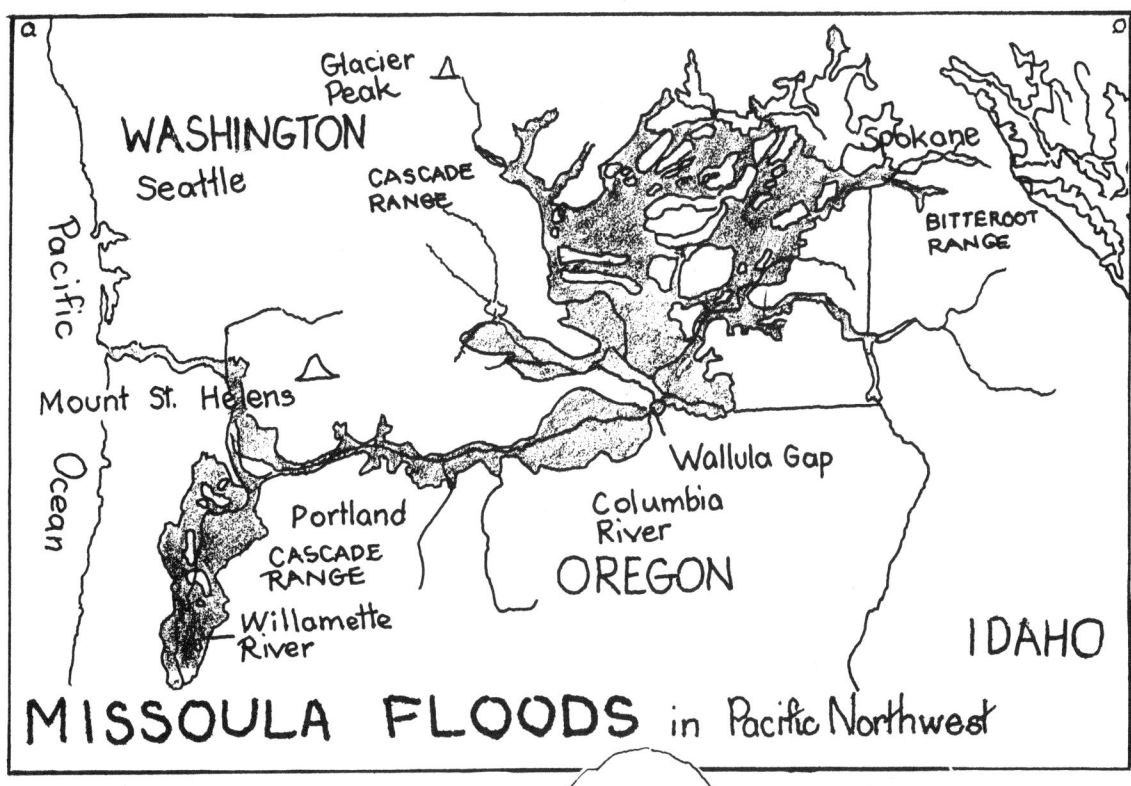

MISSOULA FLOODS in Pacific Northwest

The best geologists are said to be those who can tell the best stories! Geologists study hills, mountains, rocks, and rivers, observable pieces of the earth's history, and come up with possible explanations. Is there room for doubt? —
 Absolutely!
Disagreement? —
 You betcha!
Change in theory? —
 Could be!
My recommendation:
Dig in and study! Learn the tools of the trade, study the observable characteristics and discover reasons and explanations for all you see around you.

PLATE TECTONICS

(to the tune of "Wabash Cannonball")

The Wabash River runs in Western Ohio. The Wabash Cannonball, a train, ran between St. Louis and Detroit and is celebrated in the old song.

Con- ti- nen- tal plates of crust of thin- ner lith- o- sphere. And hea- vy oc e- a- nic plates a
The mid-At- lan- tic ridge is from di- ver- gent bound- ary plates. New crust forms and pulls apart

bove as- then- o- sphere. Tec- ton- ic plates seem all to move and the cont- in- ents to drift.
where they sep- ar- ate. But volcan- ic vents called hot spots in New Zea- land and Galapa- gos,

And in vall- eys and the o- cean floor the spread- ing caus- es rift.
Ha- wa- ii and in Ice- land, a- - nd in Yell- ow- stone.

    Tectonic plates converging, oceanic plate subducts
    Under lighter continent—bends down then it melts
    Pacific Rim's a "ring of fire": volcanoes and earthquakes
    Convergent boundaries where you'll find those subducting plates

    Tectonic plates instead may move more from side to side
    At a transform boundary, strike-slip fault; at lateral faults they slide
    California's San Andreas fault is where you'll find
    One of the examples of movement of this kind

PLATE TECTONICS

As the story of Bretz demonstrates, theories in science change—slowly and with new evidence. In 1960 another theory came into acceptance which was a revolution in geologic thought. The theory of **plate tectonics** was able to explain structures, landforms, and processes on the surface of the earth. Yet its acceptance—like that of Bretz's catastrophism—did not come quickly or easily. One man, Alfred Wegener, spent years trying to convince the scientific world that the continents move apart and push together on the crust of the earth. His theory was not accepted by the scientists of his day, and his ideas were not proven until after his death, when substantial new discoveries were made.

Alfred Wegener
1880--1930

ALFRED WEGENER

Alfred Wegener (pronounced "VEG ener") was a German interdisciplinary scientist (astronomer, meteorologist, geophysicist) who looked at "the big picture" of the earth. He built upon the observations of others, such as Francis Bacon, who observed in 1660 that the coastlines of South America and Africa seem to fit together like pieces of a jigsaw puzzle. Specifically, Wegener was able to see other relationships between these continents, including:

1. similar fossils (remains or imprints in rock) of plants and animals, which gave evidence of similar climates in ancient times;
2. identical rock layers where the two continents would have been joined.

Using these observations along with other data as evidence, he developed and introduced a theory in 1912, called **continental drift**. This theory stated that the continents were drifting, almost like icebergs floating on the ocean—away from each other and from a super continent he called **Pangaea**—and plowing into each other to form landforms such as mountain ranges.

Pangaea means "all lands."

Military exploration begun during World War II combined with present-day scientific mapping of the ocean floor have provided overwhelming evidence to support Wegener. One discovery was the Mid-Atlantic Ridge, a 12,000-mile-long (19,200 km) underwater mountain range that stretches the length of the entire Atlantic Ocean. Between its ridges lies a 24 mile-wide (38 km) valley where the plates are moving away from each other.

Wegener's theory has since been revised— no longer are continents considered to be merely drifting on the surface of the earth; instead, they are parts of large **plates** of crust that are riding atop hot partially molten (melted) rock. However, his observations helped form the modern understanding and theory of plate tectonics, which explains much about how mountains are formed, how volcanoes are created, and what causes earthquakes.

Mapping of the ocean floor began during World War II with the use of submarines. An unexpected result was evidence supporting Wegener's theory of sea floor spreading.

PARTS OF THE EARTH

Before looking at the surface of the earth and how landforms are created by tectonic plates, we need to understand the interior of the earth. It has different layers, each with its own unique physical and chemical characteristics.

1-lithosphere—the crust and tectonic plates, also includes the uppermost part of the **mantle** (the layer between the crust and the core).
2-asthenosphere—partially liquid mantle below the lithosphere.
3-core—the outer core which is liquid and the inner core which is solid.

The earth's outermost shell is the **lithosphere** (from the Greek *lithos*, meaning "stone" and *sphere*, meaning "globe"). The crust is attached to plates of the top of the mantle and make up the lithosphere. This "stony globe" is hard and rocky in nature. The lithosphere rides on the next layer, the **asthenosphere** (from the Greek *asthenes*, meaning "weak"). The contents of the "weak globe" contains a small amount of **magma**, hot molten rock, and rock creating a consistency said to be like chunky, thick applesauce. The outer **core** is thought to be made of molten iron, and the inner core, solid and dense iron.

Journey to the Center of the Earth was a popular novel in the late 1800s about adventurers traveling toward the core of the earth. If you really could go to the center, you would first travel through 3 to 60 miles (4.8 to 96 km) of lithosphere (includes the crust), 1,800 miles (2,880 km) of mantle (includes lithosphere and asthenosphere), 1,400 miles (2,240 km) of outer core, and finally 750 miles (1,200 km) of solid inner core to get to the very middle of the earth—4,000 miles (6400 km)!

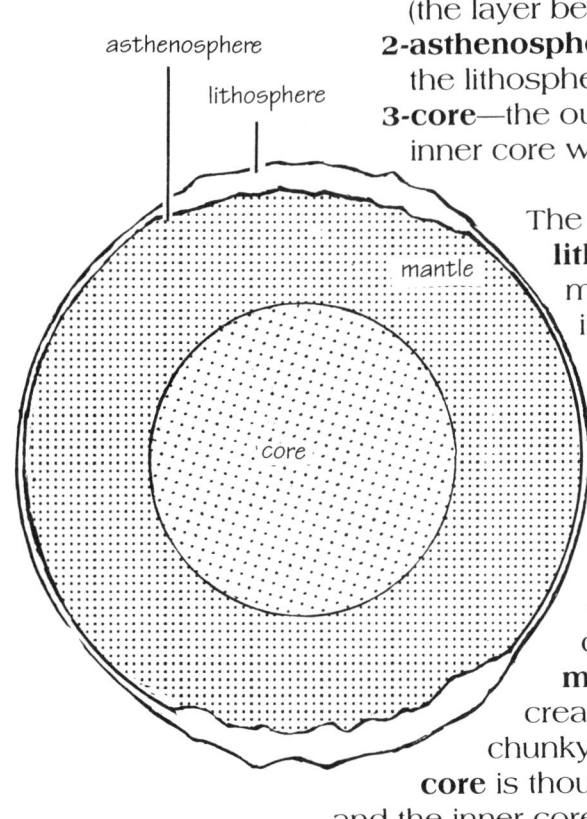

Parts of the earth: asthenosphere, lithosphere, mantle, core

TECTONIC PLATES

The lithosphere is made up of approximately twelve large and numerous smaller plates. These hard, cold masses of rock sit on the hot, partially molten rock of the asthenosphere which act like ball bearings for the plates' movements. The two kinds of plates include:

- **1-oceanic plates**: relatively thin, 3 to 5 miles (5 to 8 km) thick located beneath the ocean.
- **2-continental plates**: 25 to 40 miles (40 to 65 km) thick on which the land rides.

These plates move because the mantle is thought to have circulating currents, somewhat like water boiling in a pot, causing the plates to drift slowly—very slowly, at 0.79–6.7 inches (2–17 cm) per year. The movement of these sections, or plates, may result in collisions that build up the earth's crust. The term **tectonic plates** comes from the Greek term *tect* meaning "to build" (as in "architect"—chief builder). The science of plate tectonics is the study of these broken pieces, especially how they interact at their edges, changing the shape and surface of the earth. The plate boundaries are the "construction zones" of the earth.

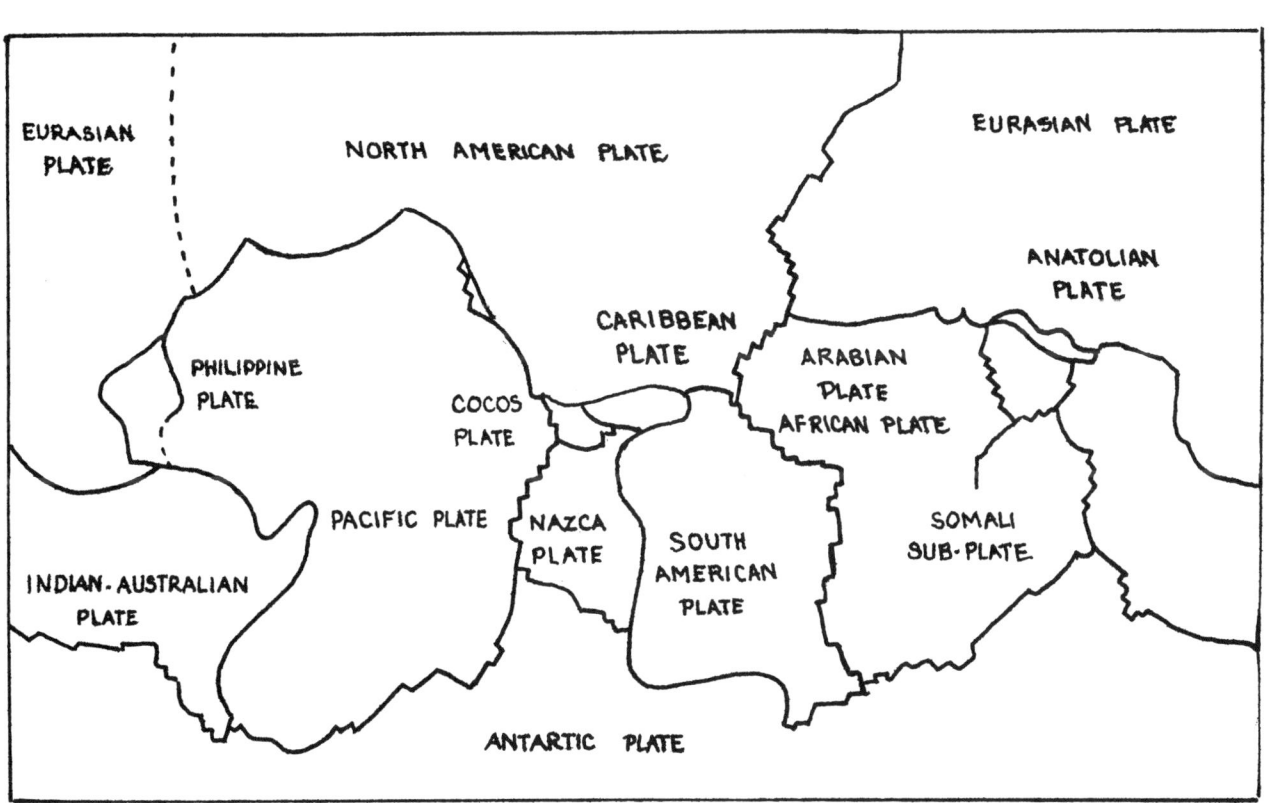

Major tectonic plates

PLATE BOUNDARIES

Three types of boundaries where the plates meet are created by three different directions of movement and result in various crust building activities.

> A corny geology rhyme:
> Divergent boundaries
>   are where plates divide
> Convergent boundaries
>   are where they collide
> Transform boundaries
>   are where they slide

1. At **divergent boundaries** plates are moving farther apart (*di* means "divide" or "apart").
2. At **convergent boundaries** plates are moving in and squishing together (*con* means "together").
3. At **transform boundaries** plates slide sideways against each other.

Divergent boundaries, or **divergent zones**, are where plates are moving apart in opposite directions and so are sometimes called **spreading zones** or **ridges**. In a process called **sea floor spreading**, magma pushes up between plates causing earthquakes, creating volcanoes, and forming new crust as it cools. There are many ocean ridges—most divergent boundaries are found in the ocean where they create wide **rift valleys** bordered by tall volcanic mountain ranges. The Mid-Atlantic Ridge (mentioned on page 14), is formed by the diverging Eurasian and American Plates. It is only one section of a mid-ocean ridge, extending 40,000 miles (64,360 km) from the Arctic Ocean to the Atlantic Ocean, around Africa, Asia, and Australia, and under the Pacific Ocean to the west coast of North America. Its highest point is 13,800 feet (4206 m) above the base ocean depth.[1]

> An example of a divergent boundary on a continent is the 2,500 mile (4,022 km) long Great Rift Valley in East Africa.

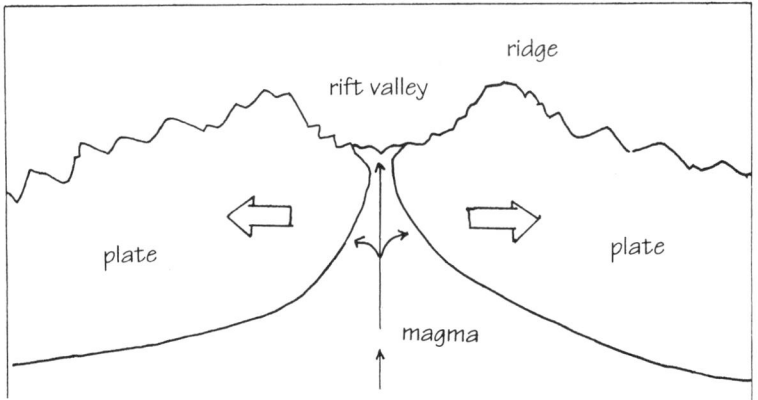

Divergent boundaries

Convergent boundaries are where plates meet, so they are sometimes referred to as collision boundaries. They create a **subduction zone** (*sub* means "under") where a heavier oceanic plate converges with a lighter continental plate, causing the heavier plate to bend, or subduct, under the continental plate. Deep **trenches** (long, narrow, deep basins) are created as the oceanic plate sinks into the mantle and melts. One of the deepest trenches is the 6.5-mile-deep (10.5 km) Mariana Trench, where the Pacific Plate meets the Philippine Plate!

Plate Tectonics

Subduction zones are areas of intense geological activity—places where there are many earthquakes and volcanoes. Mt. St. Helens, in the state of Washington, erupted in 1980. It was formed by the convergence of the Juan de Fuca and the North American Plates. This volcano is only one of over 500 found in the "**ring of fire**," a string of volcanoes and **faults** (cracks in the earth where there is movement) surrounding the 30,000-mile-long (48,300 km) rim of Pacific Ocean where the Pacific Plate, and its smaller adjacent plates (including the Juan de Fuca) meet continental plates.

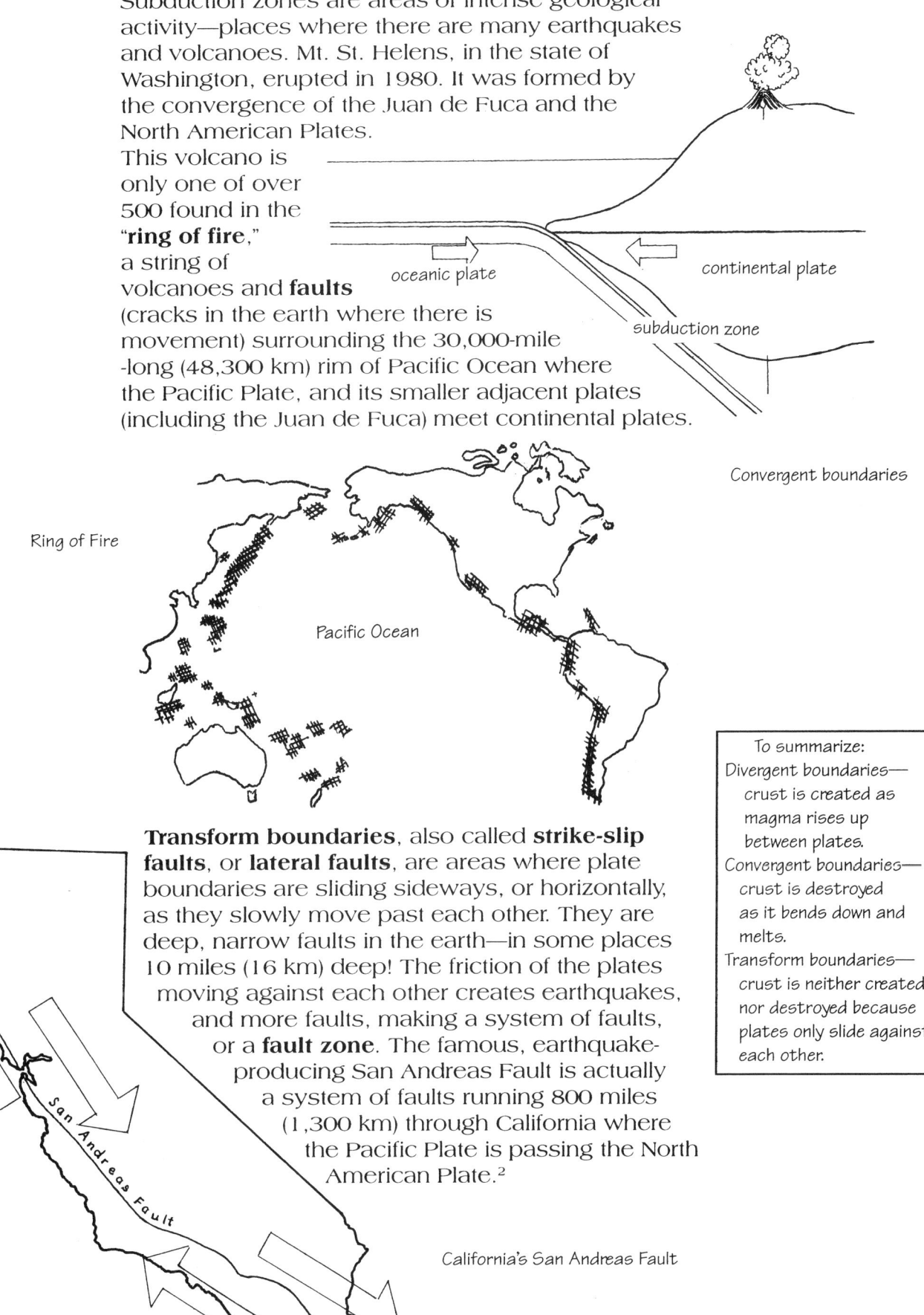

Convergent boundaries

Ring of Fire

Pacific Ocean

Transform boundaries, also called **strike-slip faults**, or **lateral faults**, are areas where plate boundaries are sliding sideways, or horizontally, as they slowly move past each other. They are deep, narrow faults in the earth—in some places 10 miles (16 km) deep! The friction of the plates moving against each other creates earthquakes, and more faults, making a system of faults, or a **fault zone**. The famous, earthquake-producing San Andreas Fault is actually a system of faults running 800 miles (1,300 km) through California where the Pacific Plate is passing the North American Plate.[2]

California's San Andreas Fault

> To summarize:
> Divergent boundaries—crust is created as magma rises up between plates.
> Convergent boundaries—crust is destroyed as it bends down and melts.
> Transform boundaries—crust is neither created nor destroyed because plates only slide against each other.

FOLDING, FAULTING, AND INTRUSION

(to the tune of "Washington Post" March)
Probably the most popular patriotic march written by John Philip Sousa just before the end of the 19th century.

Although many mountains are volcanoes, there are forces that will change the crust of Earth into a range of mountains which are not volcanic. Tectonic forces cause the folding and the faulting of the crust. And with faults there are three main kinds. Direction of force will vary of course. The foot-wall and the hanging wall, if moving apart the tension causes a normal fault. Intrusive igneous rock bodies, where magma underground slowly cooled into rock can be as large as an entire mountain system (in batholiths) or smaller in a stock.

Part A: And when these forces do cause folding
Creating A-shaped anticlines and U shape of synclines
Twisting, folding, squeezing, pulling
Movement that's along a fracture is a fault and not a joint

Part B: But if the force is the other way
Compression will push, a fault in reverse
But then a shear or a sideways force
Can make a strike-slip fault like San Andreas

Part C: A dike is magma that hardens after cutting 'cross the layers
A sill is where magma squeezed between
And a laccolith is like a sill except that it's thicker
Pegmatite is where large minerals are seen

Part C: So now you know some structural mountain forms
From folding, faulting, intrusion, is uplift
But if the landforms are formed instead from erosion
Dissected mountains are what is left

FOLDING, FAULTING, AND INTRUSION

Tectonic plates, pulling away and colliding against each other, create everything from mountain ranges to valleys. These actions cause the crust to stretch, bend, and break, changing its original shape in a process called **deformation**. As plates move farther away from each other at one location, they move closer together somewhere else. The different **forces**, or **pressures**, result from plate movements and correspond to the three different types of boundaries.

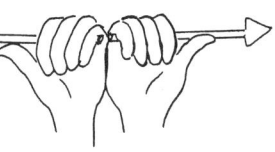

Direction of tension

1-**Tension**, or **extension**, results when plates thin or separate at divergent boundaries.
2-**Compression** occurs when plates crunch together at convergent boundaries.
3-**Shear force** results when plates slide side by side at transform boundaries.

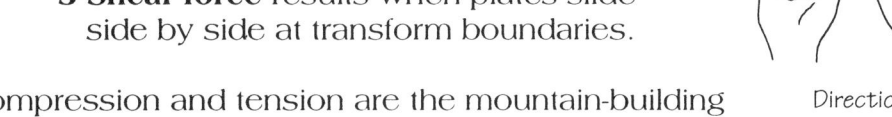

Direction of shear force

Direction of compression

Compression and tension are the mountain-building forces which form two main types of mountains, areas of **uplift** where the crust has been raised.
1-folding—the crust buckles but doesn't break as the plates converge.
2-faulting—the crust breaks and slips causing one edge to be higher than the other as the plates converge.

FOLDING

The mountain building process that creates **folded mountains** occurs when two continental plates, two huge masses of rock, (possibly thousands of square miles!), compress against one another. The crust buckles and folds in response to the pressure, creating long, tall mountain ranges. For example, when the Indian Plate converged with the Asian Plate, the crust folded into a huge series of waves creating the Himalaya Mountains—the highest mountain range in the world. Even now the Himalayas are still growing—at about 2 inches (5 cm) every year because the plates are still compressing against each other.

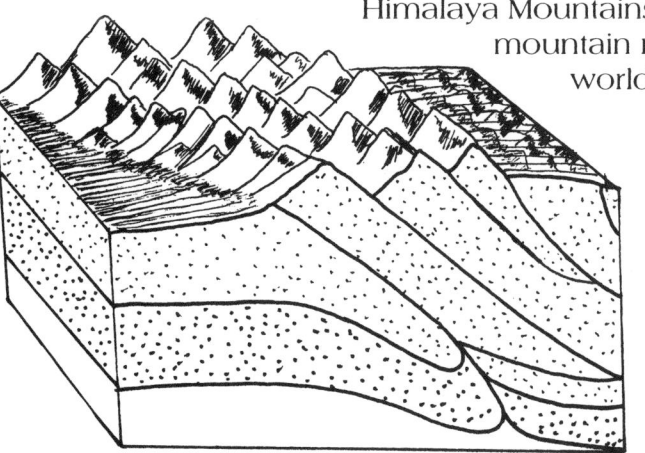

Formation of folded mountains occurs as converging plates form folds in the crust.

> Other mountain ranges formed by folding include the Appalachians and the Alps.

The compressive folding process can be so intense that sedimentary beds of older rock can be bent until they are on top of younger rock, creating an **overturned fold**. Folding creates curves and bends called:
    1-anticlines: A-shaped slopes at the rising of the fold;
    2-synclines: the lower area between anticlines.

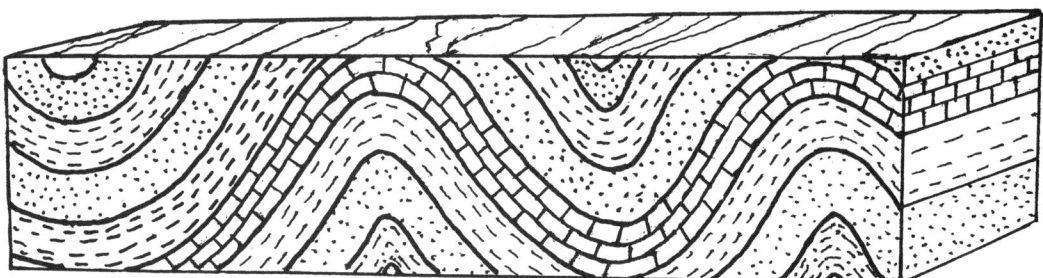

FAULTING

Other types of mountains, **fault block mountains**, are created when forces of tension or compression cause the crust to break, or **fracture**. If there is movement of the crust on either side of the fracture, it is then called a **fault**; if there is no movement it is a **joint**. Faulting causes huge chunks, or blocks, of crust to thrust up or slide down—the crust is uplifted leaving one side towering over the other. Fault block mountains may have the following characteristics:
    1- a steep slope on one side which,
    2- rises sharply from a flat valley;
    3- loose rock, or debris, called an **alluvial fan** near its base (caused from moving water, such as a **flashflood**, a severe, violent river of floodwater);
    4- a gradual slope on the other side.

Stunning examples of fault block mountains are the Panamint and Armagosa Mountains on either side of the famous Death Valley in California.

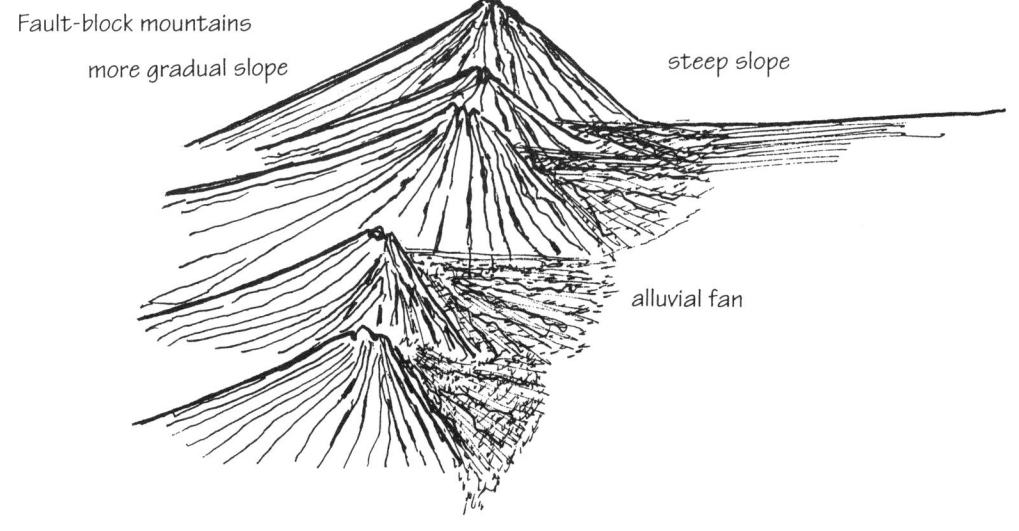

Folding, Faulting, and Intrusion

Though the mountain-building tectonic forces of tension and compression are strongest at plate edges, they can also create faulting within plates. Faults are classified by their **slip**, the direction of movement of the blocks of earth. The fault illustrated below shows the individual features that geologists study in detail to understand how particular mountains are formed.

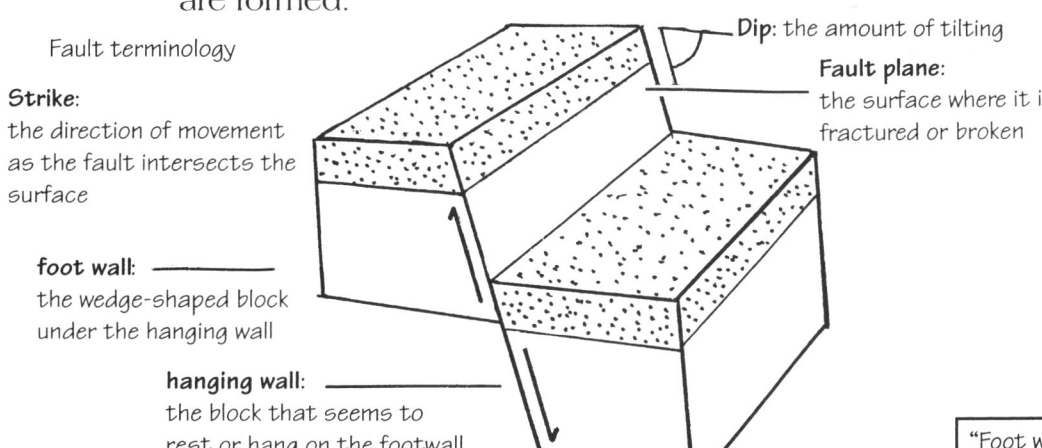

Fault terminology

Strike: the direction of movement as the fault intersects the surface

Dip: the amount of tilting

Fault plane: the surface where it is fractured or broken

foot wall: the wedge-shaped block under the hanging wall

hanging wall: the block that seems to rest or hang on the footwall

"Foot wall" and "hanging wall" are terms given by miners of long ago. The foot wall is where they walked and the hanging wall is the block of earth that hung over their heads in reverse faults.

Common types of faults include:

More fault-created landforms:
1- **grabens** are valleys formed as extension drops an area between two normal faults.
2- **horsts** are formed as extension leaves an area between two parallel normal faults higher than the adjacent valley.

1- **normal fault**: crust moves along the fault plane as it is pulled apart by **tension**, or **extension**. The crust above the fault plane moves down, causing a lengthening of the crust.
2- **reverse fault**: crust also moves along the fault plane when it is pushed together by **compression**. The crust above the fault plane moves up, causing a shortening of the crust.
3- **strike-slip faults**: crust is neither lengthened nor shortened but moves parallel to the strike by **shear force**. The crust slides laterally, or side by side.

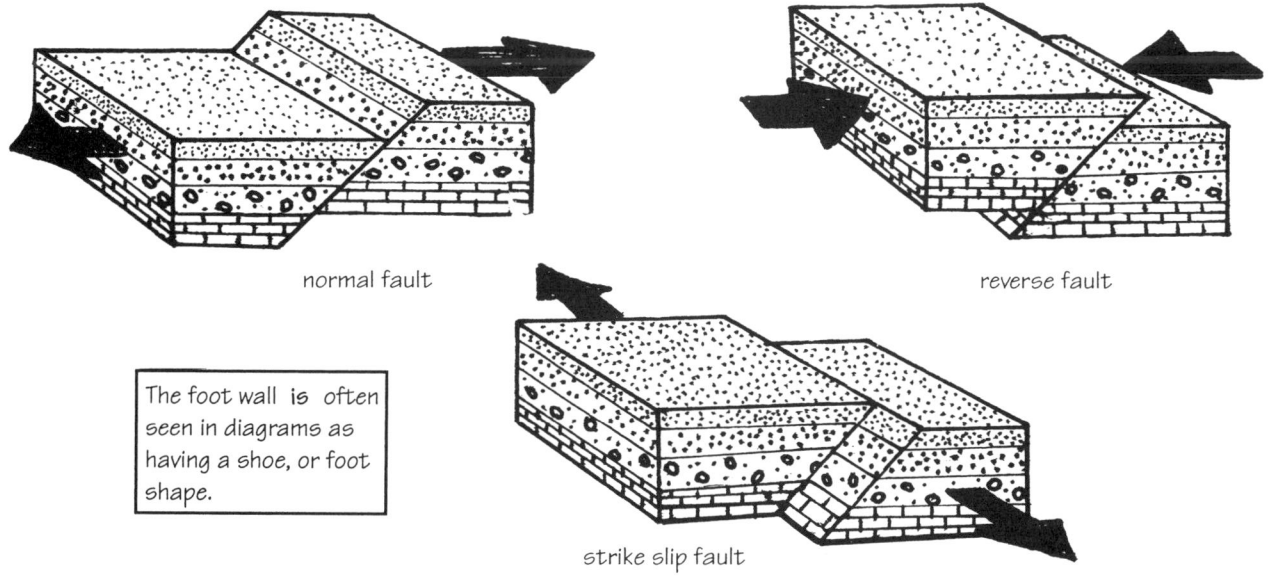

normal fault

reverse fault

strike slip fault

The foot wall is often seen in diagrams as having a shoe, or foot shape.

Another type of mountain may have gentle folding and faulting, but **dissected mountains** are formed by the process of **erosion** (studied in detail beginning on page 62) of a **plateau**, a large, relatively flat, raised area—also called a tableland. Dissected mountains are formed as streams and rivers cut through the ground—rocks and rock particles are carried downhill and laid down. The once flat terrain is transformed into hill tops and valleys—all that remains after streams and rivers have carved their paths through the landscape. The Boston Mountains of Arkansas are the hill-like examples of dissected mountains.

> Two other landforms created by erosion are **mesas** and **buttes**. Like plateaus, mesas are flat, elevated areas, but they are much smaller and have steep sides. Buttes are like mesas except even smaller.

> The Grand Canyon is a plateau that has been dissected by erosion.

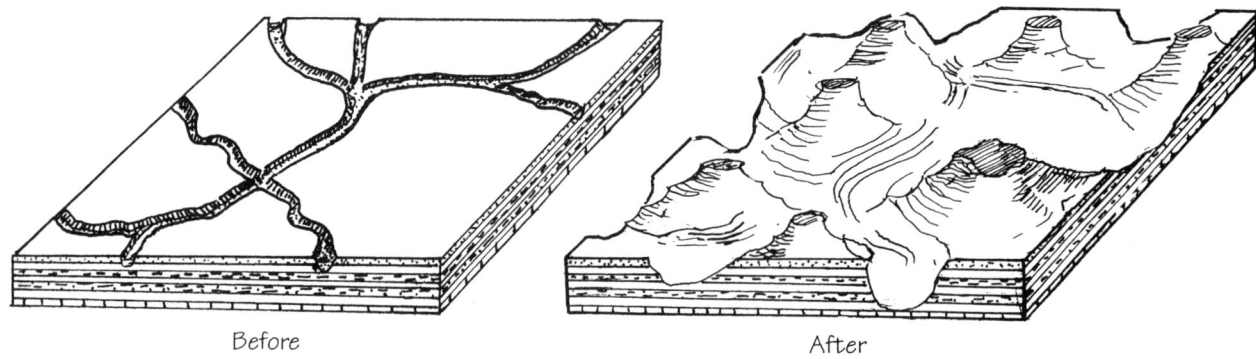

Before   After
Dissected mountains

SUMMARY
Before we move on to **intrusion**, the process of magma seeping into the crust, let's summarize the tectonic plate information of the last chapters, including directions of force and what they create as builders on the surface crust.

	Tension or Extension	Compression	Shear
Plate interaction	spreading or thinning pulling apart	pushing, squeezing folding together	sliding against each other
Common location	divergent boundaries or divergent zones	convergent boundaries or subduction zones	transform boundaries
Fault type created	normal	reverse	strike slip
Landforms and geologic features	fault block mountains sea floor spreading rift valleys, oceanic ridges volcanoes	folded mountains fault block mountains ocean trenches volcanoes	altered landscapes, as in rivers abruptly changing course, linear lakes
Examples of landforms	Mid-Atlantic Ridge Great Rift Valley Death Valley's Basin and Range	Himalaya Mountains Appalachian Mountains Alps	San Andreas Fault
Earthquake type	shallow, moderate	deep, large	shallow, large

Folding, Faulting, and Intrusion

INTRUSION

The term **intrusion** describes what magma actually does: it "intrudes" into the existing rock, known as **country rock**. When the intruding magma cools and hardens, it becomes igneous rock (see page 43), and creates particular types of landforms, geologic structures, and rock forms.

Batholiths and other igneous formations, such as **volcanic necks** (exposed inner parts of volcanoes) can be seen when the softer rock above and around them is worn down and carried away by wind or water.

1-**Batholiths**, huge masses of hardened intruded magma are considered the "roots" or the base of mountains and reach deep into the earth. Parts of the Sierra Nevadas in California are batholiths which have been exposed as some of the mountain has been worn away.

2-**Stocks** are like batholiths but smaller.

3-**Sills** are created where magma squeezes between rock layers, then hardens to form horizontal layers.

4-**Laccoliths** are like sills but with a lot more magma which squeezes between rock layers to form the shape of flattened mushrooms. Laccoliths are thicker than sills; so thick that they lift up the overlying rock.

5-**Dikes** cut up through and across rock layers.

6-**Pegmatites** are pockets of large crystals sometimes found in dikes and sills formed from leftover water and elements in the melted rock as it slowly cools. The crystals themselves point inward and can be quite long, up to 42 feet (13 m)![3]

The Mittens, in Monument Valley, Utah and Arizonia are volcanic necks

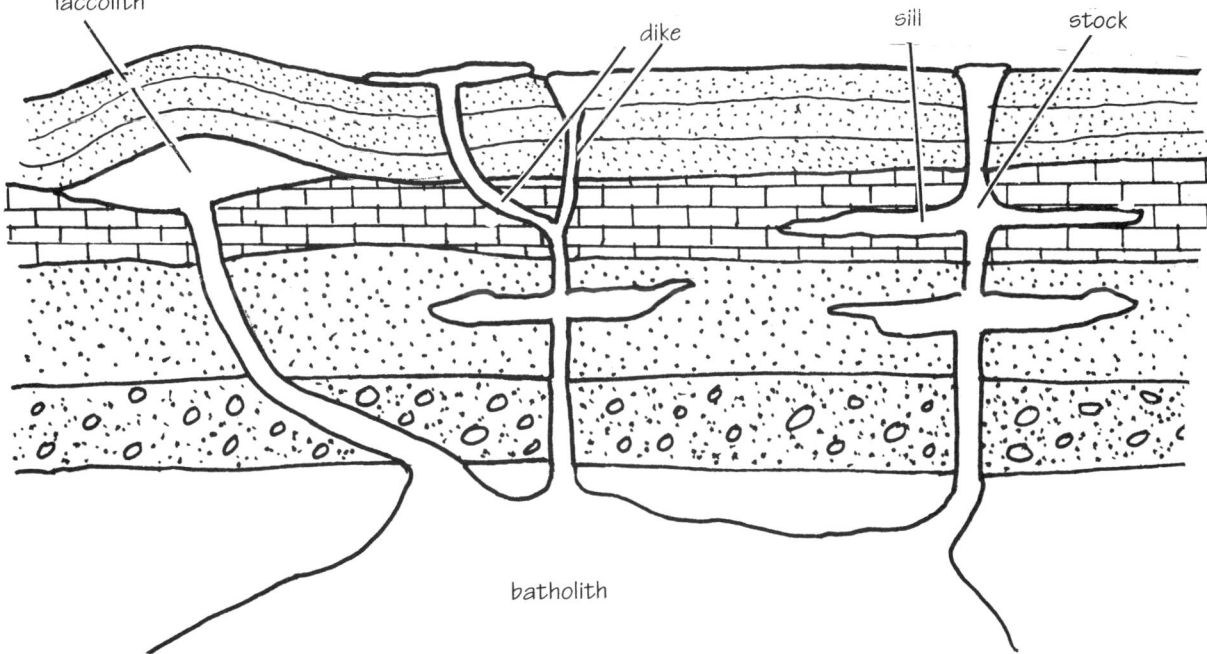

Intrusion of magma into the surface crust

EARTHQUAKES

(to the tune of "Little 'Baccy Box")
A popular old mountain melody that has been set with a number of humorous lyrics—from chickens with wooden legs, to a young man's difficulties courting a girl with many pestering siblings.

Plates converging, all the forces and the stress
From the pushing and the sliding,
    all the tension you may guess
Land begins to shake 'til releasing makes it stop
Focus in the mantle, epicenter's on the top
Focus in the mantle, epicenter's on the top

Seismographs at the seismic stations
Seismograms record the wave vibrations
Measure magnitude or intensity
Relative amount of released energy
Richter Scales and amount of energy

Deep primary waves are longitudinal
They stretch and compress and
    they push and they pull
Secondary—s waves, slower don't you know
Side-to-side or up or down they go
Side-to-side or up or down they go

And surface waves all along the surface roll
Together with the s waves they really take a toll
Shaking buildings down and shaking everything
Blame these kinds of waves
    for all trouble that they bring
Blame these kinds of waves
    for all trouble that they bring

EARTHQUAKES

An **earthquake** is the sudden shaking of the earth caused by slipping of the crust. Earthquakes are produced by the movement of tectonic plates at their boundaries such as at ocean ridges, subduction zones, transform boundaries. Earthquakes are generated at faults when plates or blocks of crust break free and release stress at their edges, where bulges and protrusions grind against one another.

Earthquakes are also caused by the underground movement of magma near volcanoes.

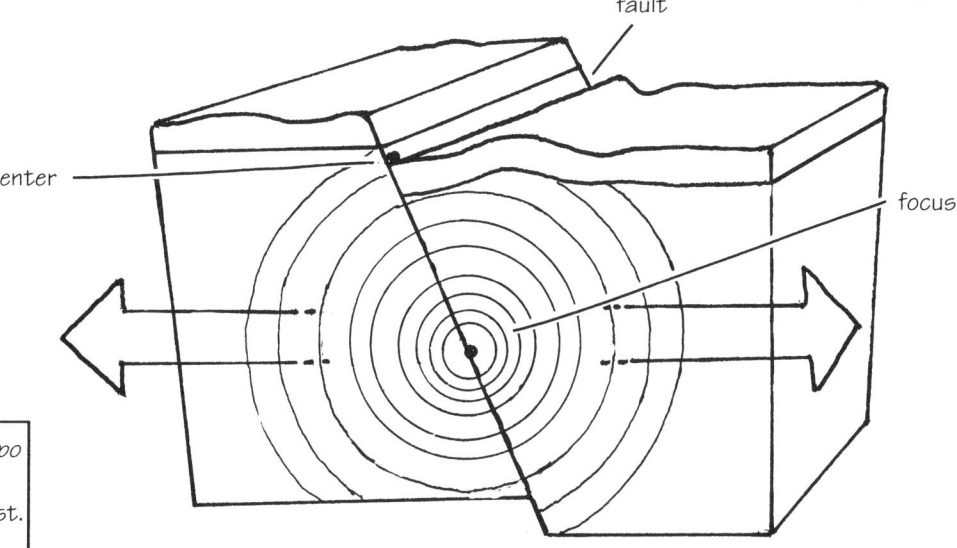

Hypocenter—*hypo* means "under"—under the crust.
Epicenter—*epi* means "up on"—upon the crust.

LOCATION

The **focus**, or **hypocenter**, which is the source of the earthquake, can be anywhere from 10 to 500 miles (16 to 800 km) underground, in the lower crust or upper part of the mantle—the lithosphere. Some of the largest earthquakes have occurred deep in the earth at subduction zones. The **epicenter** of the earthquake is on the crust's surface, directly above the focus.

SEISMIC WAVES

Seismic waves travel upwards from the focus to the epicenter and along the surface crust in much the same way water ripples when a stone is dropped in a pool. These seismic wave vibrations are of distinct types and can be detected by a scientific instrument called a **seismograph** that records their movements on printed **seismograms**.

Seismographs make seismograms

Water waves radiate outward from where a stone is dropped, similar to the way seismic waves radiate from the focus of an earthquake.

The most common seismic waves are listed.

1-**primary waves**, or **p waves**, the first waves to reach the seismograph, move by compression as they push and pull—like a slinky toy. P waves are a type of **body wave**, which means they travel inside the earth, upward toward the surface crust.

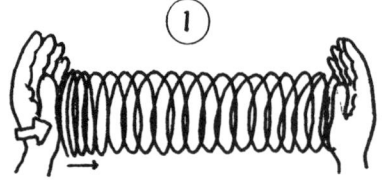
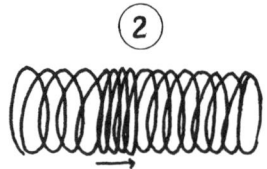
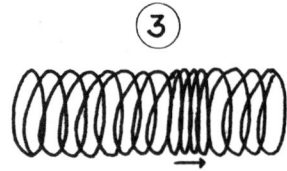

primary or p waves

2-**secondary waves**, or **s waves**, the second waves to reach the seismograph, move up and down, and back and forth the way a rope moves. They are also body waves traveling inside the earth away from the focus, but they stop when they come to a liquid, such as magma. These waves may also be called **shear waves** because they move through rock material at right angles.

secondary or s waves

3-**surface waves** are altogether different from the p and s waves because they are not body waves that travel in the earth, but, as their name implies, travel within the surface crust. Surface waves, along with s waves, cause most of the earthquake damage.

surface waves

Love and Rayleigh waves are surface waves, they move up and down and back and forth.

Seismologists, scientists who study earthquakes, use these waves to measure the strength, or **magnitude**, of an earthquake on a seismograph. The **Richter Scale** uses this information in a mathematical formula to compare the size of different earthquakes. The weakest earthquakes have a magnitude of 1 and the

The Richter scale is not a machine, but a number derived from a calculation to compare sizes of earthquakes.

strongest, a magnitude of 9; each unit is ten times that of the last unit. Earthquakes we can feel begin at magnitudes of 2.5 to 3 and there are literally hundreds of smaller earthquakes we never notice!

One of the strongest earthquakes ever recorded had a magnitude of 9.2 at Prince William Sound in Alaska on March 28, 1964.

Another way to measure the strength of an earthquake, called the **Mercali Scale**, is based on how much damage an earthquake causes. Roman Numerals denote the increasing destruction from the least to the most.

I Most people do not notice, animals may be uneasy, can be detected by a seismograph.
II Hanging objects sway back and forth.
III Many people feel the movement, parked cars may rock.
IV Doors, windows, and shelves may rattle, people indoors can feel movement.
V Light furniture moves, pictures fall off walls, objects fall from shelves.
VI Nearly everyone feels movement, light fixtures fall over, windows may crack.
VII Some people fall over, walls may crack.
VIII Heavy furniture falls over, some walls crumble.
IX Many people panic, some buildings collapse, dams crack.
X Railroad lines are bent, most buildings are damaged.
XI Bridges collapse, buried pipes break, most buildings collapse.
XII All man-made structures are destroyed.[4]

seismic gap
seismic hazard
seismicity
seismic moment
seismic reflection
seismogenic
seismometer
seismology
seismograph
seismogram
seismologist
seismic waves
seismic zone
seismic activity

OTHER SEISMIC EVENTS
Before the main earthquake there are usually **foreshocks**, smaller earthquakes in the same location either days or weeks before the main earthquake. **Aftershocks**, also smaller earthquakes, may occur hours or weeks after the main earthquake.

Seismo means "shaking," and it seems to be found everywhere, from Biblical accounts of earthquakes to corny attempts at humor:

Tsunamis, or **seismic sea waves**, are huge destructive waves caused by the shifting and resettling of the ocean floor after an earthquake or volcanic eruption. A tsunami may not be observable on the open ocean, but it may be 90 feet (30 m) high when it reaches land!

"In a seismic zone, during a seismic moment following seismic activity, seismologists studying seismology used their seismic meter and seismograph to detect the seismic waves to record them on a seismogram so they could try to anticipate the next seismic hazard at the seismic zone."

VOLCANOES

(to the tune of "The Minstrel Boy")
This melody is from another beautiful and gentle traditional Irish tune.

Mafic magma features they include
Flowing lava, high temperature is fluid
Basalt of lava fountains, falls, and tubes
Cinder cones, or most common—shield volcanoes
    Of flowing lava the main kinds are two:
    Pahoehoe lava is the one that's smooth
    And aa is so rough and broken
    Coming from Mauna Loa, the volcano

Felsic magma temperature is low
Make the pyroclastic flow from the volcanoes
And with intermediate of flow and ash
Make composite, or the stratovolcanoes
    Fissure eruptions may form plateaus
    Gasses and steam come from the fumaroles
    Lava, ashes, when they're mixed and blown:
    Spew from craters and vents of the volcanoes

VOLCANOES

> Geologists use seismographs, not only to record earthquakes, but also to predict volcanic eruptions. Small earthquakes are recorded as lava moves underground.

Like earthquakes, volcanic activity is usually near plate boundaries. But it is surprising to realize that 80 percent of it takes place at divergent boundaries in the ocean! On land, usually at convergent boundaries with subduction zones, volcanoes are known for their destructive power: they flatten villages, towns, and cities; cause massive floods and mass movements; and blow over many square miles of forest. But volcanoes are also major builders of new crust and landforms all over the world. Throughout time, volcanic eruptions have been building everything from mountain ranges on continents, such as those which border the Great Rift Valley in Eastern Africa, to islands in the middle of the oceans, such as the eight islands that make up the state of Hawaii.

Different types of volcanoes

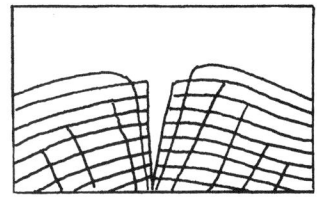
shield volcano

cinder cone

> The term *volcano* can refer either to the vent which emits lava, or the mountain formed around the vent.

A volcano's destructive and constructive activities are the result of molten rock, or **magma**, from the mantle finding a weak place in the surface crust. Magma, along with steam and gasses, erupts though a **vent**, and the **volcano** itself is the mountain formed by the eruption. When magma spills, or explodes, onto the surface, it is then called **lava**. The term *lava* can refer either to the fluid **lava flow**, or to the hardened lava particles and fragments called **pyroclastics** blown out of a volcano. The eruption often creates a **cone** and may form a bowl-shaped **crater** in the volcano. During a huge eruption, the cone may even collapse on itself, creating a large **caldera**, which is how Crater Lake in Oregon was formed.

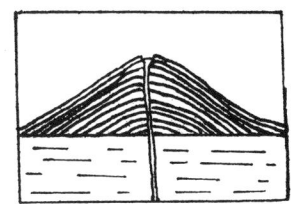
composite or stratovolcano

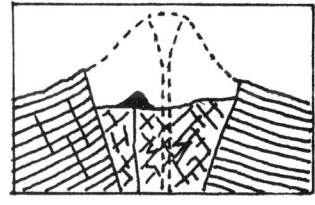
caldera

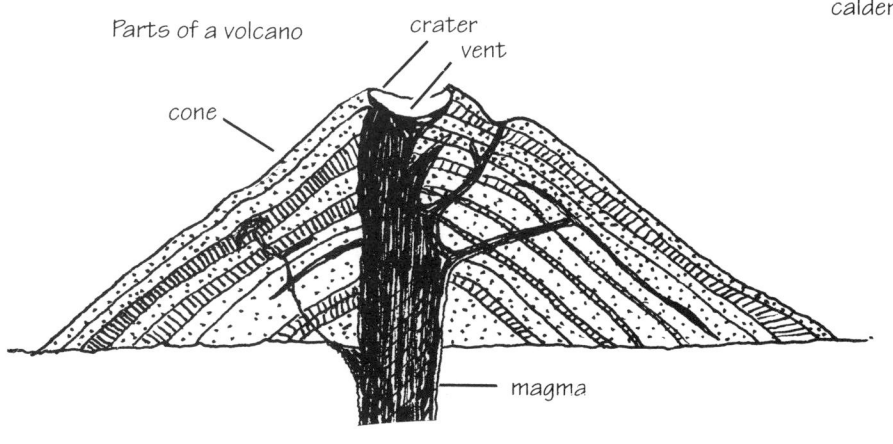
Parts of a volcano

CLASSIFICATION

Each volcano is unique, creating its own type of eruptions and landform shapes.—They do not all look like the classic conical-shape mountain. Unlike plants or mammals, whose distinct, observable features define their biological classification, volcanic eruptions do not fit into neat and tidy categories.—They often have more than one vent, more than one eruption, and different kinds of lava at various times. But in general, volcanoes are identified and classified by the following characteristics:

    1-lava composition, the chemicals, or ingredients, and their amounts in the lava.
    2-shape, or the structure, of the volcano.
    3-type of eruption, whether it is explosive or quiet.

> Volcano classification can be based on the types of eruption at the following places of historic volcanic activity, from least to most explosive:
> Icelandic
> Hawaiian
> Strombolian
> Vulcanian
> Pelean

When geologists study a lava's chemical composition, they identify the amount of one particularly important **element** (a basic substance that cannot be separated) called **silicon**. It is found in the crust often combined with other elements, such as oxygen, to make **mineral compounds** (discussed further on page 37) called **silicates**. Silicates are the most common mineral compounds in the crust. They cause a lava to be **viscous**, or sticky and thick. The amount of silicates determines a lava's type, which determines a volcano's characteristics, such as the type of eruption, and the resulting shape. There are three kinds of lava.

    1-Mafic lava is low in silicates and, therefore a fluid lava. (The term *mafic* is formed by *Ma* from the element "magnesium," and *Fic* from Ferric for "iron.")
    2-Felsic lava is high in silicates and so is a sticky, thick, and hard lava. (The term *felsic* is derived from *fel* for "feldspar" and *sic* for "silica," a kind of silicate.)
    3-Intermediate lava has more silicates than mafic but less than felsic, so it has characteristics between the two.

> Volcano classification can be based on the shape of the volcano. Just a few of the common types, illustrated on the previous page are.
>   Shield Volcano
>   Composite Volcano
>   Cinder Cones
>   Calderas
> But,
> *Volcanoes of the World* states there are really 26 different kinds!—
>   699 stratovolcanoes,
>   164 shield volcanoes,
>   105 submarine volcanoes,
>   92 monogenetic fields (which may contain hundreds of vents, some of which may be cinder cones),
>   87 isolated cinder cones,
>   83 caldera complexes.
> Another 20 types of volcanoes make up the remainder of the volcanoes on Earth.[5]

Mafic lava—oozes
Felsic lava—KABOOMS
Intermediate lava—kabooms [6]

Test question:
Why do these three lavas erupt so differently?

Volcanoes

MAFIC LAVA

Mafic (MAY fic) **lava**, also called **basalt**, is hot, dark (from magnesium), and heavy (from iron), but has a low **viscosity** (from few silicates). It is thin, runny, and spreads out quickly. It doesn't so much explode, as it oozes or sprays from the vent, creating fluid lava flows. The volcano shape created is most often a **shield volcano**, with a wide base and gentle slope, because the lava runs far and wide.

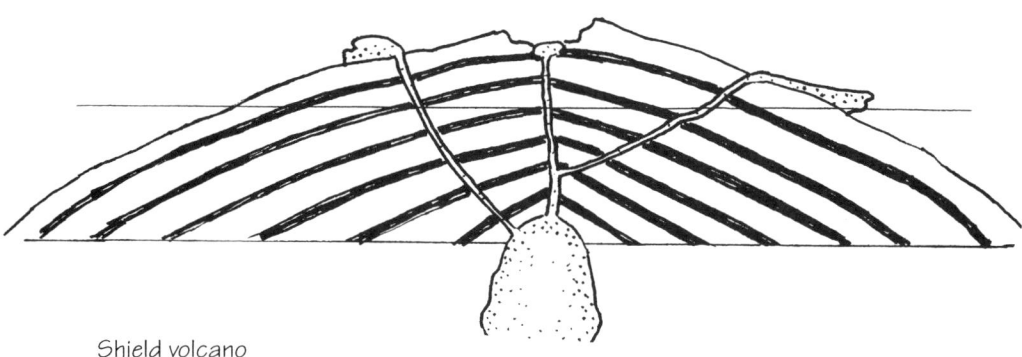
Shield volcano

The three main types of mafic lava are:
1. **pahoehoe** (pahoyhoy) is ropy and smooth—after it hardens. You can walk on it with bare feet and it won't be too uncomfortable.
2. **aa** (ah-ah) is rough, jagged, and broken—keep your shoes on or it will cut your feet!
3. **pillow lava** is formed when lava runs into water, or is erupted underwater, and rapidly cools making pillow shapes.

Three types of mafic lava

pahoehoe

aa

pillow

As the sound of their names suggests, these lavas are found in Hawaii. To be more specific, these lavas actually built the Hawaiian islands!—And many others like them in the Pacific Ocean. These **volcanic islands** were formed as the Pacific Plate moved over a **hotspot**, a deep mantle plume, sometimes called a blowtorch, allowing magma to escape to the crust. But volcanic activity is still happening, Kilauea is the most active volcano in the world, and nearby Mauna Loa, on the Big Island of Hawaii, is the largest volcano in world, when measured from the ocean floor.

The Hawaiian Islands are thought to have formed when the Pacific Plate moved over a hotspot, creating one volcano (volcanic island) after another.

Mafic lava can build landforms and geologic features such as:

1. **cinder cones**: smaller eruptions on the side of a shield volcano, built by cinders, or scoria.
2. **lava fields**: wide areas of hardened lava that can look like fields of plowed soil if made of aa.
3. **lava tubes**: caves formed after hot lava flowed through tunnels, or **conduits**.
4. **lava plateaus**: level land areas formed by **fissure eruptions**, huge cracks where mafic lava oozed out over a wide area.
5. **lava fountains**: lava combined with gasses that can shoot up to 1000 feet in the air!
6. **lava falls**: lava that slides over a cliff like a waterfall.

Scoria, or **cinders**, are tiny fragments of mafic or intermediate lava that hardened after being blown from shield or stratovolcanoes. The term *scorua* comes from the Greek for "refuse" or "trash," but scoria is put to good use for building roads in volcanic areas.

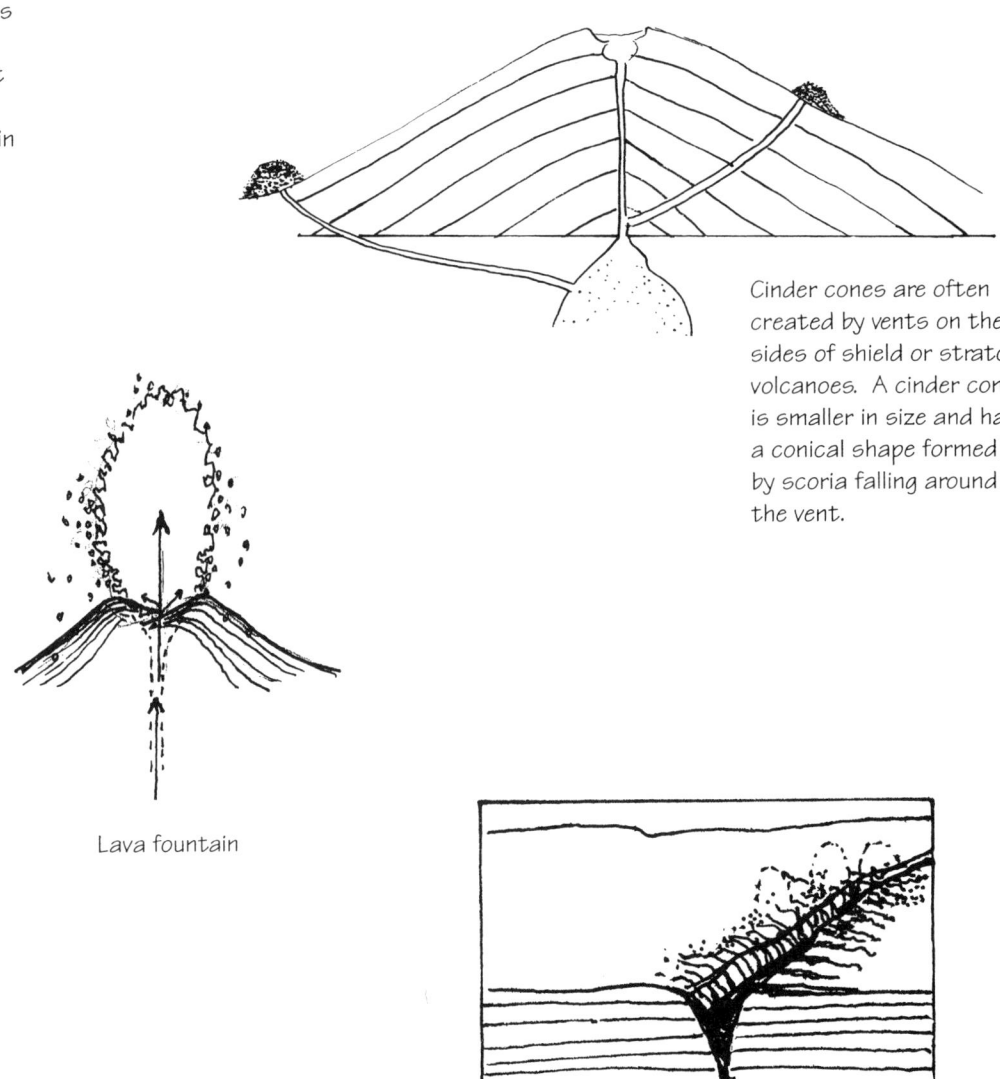

Cinder cones are often created by vents on the sides of shield or stratovolcanoes. A cinder cone is smaller in size and has a conical shape formed by scoria falling around the vent.

Lava fountain

Fissure eruptions of mafic lava created the Columbia Plateau in the Pacific Northwest covering 63,000 sq. miles (101,367 sq. km) and in some places is up to 6,000 feet (1,829 m) thick!

Volcanoes

> The largest eruption ever recorded was that of Krakatoa in Indonesia, which erupted in 1883. It shot ash up an estimated 50 miles, which spread it around the world and took 2 years to settle. It destroyed two-thirds of the island as it caved in on itself, creating a caldera five miles across. Krakatoa also caused a tsunami 121 feet (36 m) at the crest, or top, of the wave![7]

FELSIC LAVA

Felsic lava (**rhyolite**) contains a mineral called feldspar, which gives it a characteristic light color. It is also composed of large amounts of silicate minerals, which make this relatively cooler lava thick and viscous—very different from mafic lava. The high viscosity of felsic magma causes a build-up of pressure in a volcano because lava can't move fast enough to relieve the force behind it. This makes for HUGE eruptions that eject pyroclastics (*pyro* means "fire," *clastics* means "broken" and "fragments"), pieces of lava that explode out of the vent. Eruptions of felsic lava is thought to have created the volcanic area of Yellowstone National Park. Felsic lava causes explosive eruptions when it layers with mafic and intermediate lavas to form **composite**, or **stratovolcanoes**, the most common type of volcano.

Pumice, "the featherweight of rocks," is made of felsic lava and lots of gas bubbles, allowing it to float on water!

Felsic lava produces pyroclastics of numerous shapes, forms, and sizes, collectively called **tephra**. A few of the most common pyroclastics are:
1. **ash**: gritty dust particles.
2. **tuff**: compressed and hardened ash.
3. **lava blocks**, or **volcanic blocks**: huge angular pieces of hardened lava that can be as big as a car, weigh two tons, and be thrown two miles up into the air!
4. **lava bombs**, or **volcanic bombs**: huge pieces of hardened lava, that can be the same size as lava blocks but were thrown from the vent while still pliable, allowing them to form aerodynamic shapes while flying through the air.

ash

lava block

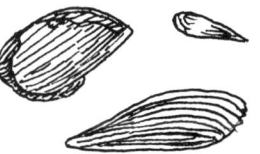

lava bomb

The dangerous **pyroclastic flow** is a hot, swiftly moving (150-mile-per-hour or 240 kmh) cloud of gas, dust and ash. The pyroclastic flow can be the most deadly aspect of a volcano. Many of the inhabitants of ancient Pompeii, Italy, were killed from the pyroclastic flow from nearby Mt. Vesuvius.

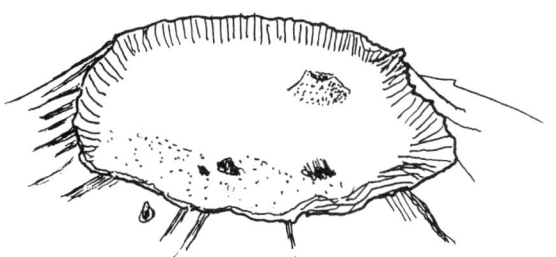

Felsic lava can create calderas, huge craters that form when the volcano collapses on itself, such as the caldera of Crater Lake, which later filled with water.

INTERMEDIATE LAVA

Intermediate lava (andesite) has characteristics, such as color and viscosity, between the thin mafic, and thick felsic lavas. It erupts in the cone-shaped stratovolcanoes which may alternately spew out the three different kinds of lava along with clouds of gas and ash. Mt. St. Helens, which blew up in 1980 after being dormant (noneruptive) for 123 years, is a modern-day example of a volcano's destructive power. The north face of the mountain collapsed then exploded into the nearby lake and river, and 150 sq. miles (240 sq. km) of forest were destroyed.

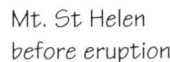
Mt. St Helen before eruption

Mt. St. Helens after eruption

Both felsic and intermediate lava can create **volcanic** or **lava domes**, which are plugs of thick lava that form over a volcanic vent before or after a major explosion.

Volcanic or lava domes:
The lava is so viscous that it squeezes out of the vent like toothpaste from its tube. The lava often forms a plug that later may get blown off by an eruption.

The chart below compares mafic and felsic lavas and the type of volcanoes where they are found. For sake of clarity, intermediate lava is not included because it shares characteristics of the other two.

lava type	eruption	emits	shapes	geological features	examples
mafic (basalt lava)	quiet (oozes)	lava flow steam, gas scoria	shield volcano fissure eruption cinder cone stratovolcano	lava tube lava plateau lava field crater	Mauna Loa, Hawaii Kilauea, Hawaii Mt. Kilimanjaro
felsic (rhyolite lava)	explosive (kabooms)	pyroclastics (tuff, blocks, bombs) pyroclastic flow (steam, gas, ash)	stratovolcano caldera	lava dome crater	Yellowstone Mt. St. Helens Mt. Vesuvius Krakatoa Crater Lake

Volcanoes

OTHER FORMS OF VOLCANIC ACTIVITY

hot springs

1. **Fumaroles** are small cracks in a volcanic region that vent gases and steam. Black Growler in Yellowstone National Park is a frequently visited fumarole.
2. **Hot springs** are springs and pools of water heated by magma in nearby crust. The water is usually full of minerals and considered to be very healthy. Hot Springs, Arkansas, has over 2 million people a year soak in the springs!
3. **Geysers** are fountains of water that shoot out of the ground at regular intervals. Water, thought to be in held underground cavities, is heated by nearby magma, then erupts as the temperature rises. Old Faithful in Yellowstone erupts every 45–90 minutes.

Geysers are so popular that there are groups of people throughout Yellowstone National Park monitoring geysers such as Old Faithful, who radio each other whenever an eruption is about to start.

Okay, so now you know that volcanoes are powerful builders. But what about their destructive nature—their ability to destroy whole cities and transform countrysides! Do you need to be afraid? Well, that depends. Do you live by one of the 600 volcanoes considered **active** (those erupting or about to erupt)? Probably not! It's more likely you live near, or maybe even on, a volcano that is **inactive** (hasn't erupted in recent history), **dormant** (hasn't erupted in a long time, but could) or **extinct** (probably won't erupt again). Basically, you won't be hearing from these last three types anytime soon. So you're safe and you may even want to visit one and hike it, bike it, ski it—the possibilities are endless!

Vulcanologist—a person who studies volcanoes.
(Vulcan was the Roman god of fire.)

MINERALS

(to the tune of "Invalid Corps")
The "Invalid Corps" was a popular parody written to shame those who would not fight for the Union during the Civil War.

Minerals are inorganic solids found in nature, with
certain properties like crystal form, cleavage, and fracture, and
hardness density or heft, color, streak and luster.
Constant composition of their chemicals' a feature.

Chorus
Minerals are inorganic solids found in nature;
Constant composition of their chemicals' a feature.

Minerals on earth are found as element or compound
And out of several thousand only twelve are commonly found
Rock made out of minerals with silicon are silicates
Other groups are oxides, carbonates, sulfates, and phosphates
Chorus

Minerals called precious metals: silver, gold, and platinum
Industrial-type metals: iron, copper, tin, titanium
Nickel, manganese, cobalt, lead, borax, molybdenum
You'd be quite surprised at all the products that contain them
Chorus

Gemstones are the minerals rare and of high value
Brilliant, durable, colorful jewelry they're made into
Diamonds, emeralds, sapphires, and rubies are examples
Turquoise, malachite, and jade—a few of opaque samples
Chorus

MINERALS

Minerals are the basic building blocks for creating all rocks. There are several thousand different kinds of minerals, but only twelve are common and only eight are **rock-formers**, including: quartz, two types of mica, two types of feldspar, olivine, pyroxene, and amphibole. These minerals make up most of the crust. Given knowledge and the right tools, anyone from miners to **mineralogists** (geologists who study minerals) can deduce what minerals comprise specific rocks. Early miners looked for minerals considered **precious metals**, such as gold and silver, or **gemstones** such as diamonds and emeralds. But all kinds of minerals have been mined and used for making everything from soap (borax) to bikes (molybdenum) to bombs (uranium).

Atoms
make
Elements
make
Minerals
make
Rocks

A few **elements**
hydrogen
oxygen
carbon
silicon
nitrogen
phosphorous
sulfur
iron
gold

A few **minerals**
(either element
or compound)
quartz
feldspar
olivine
mica

A few **rocks**
granite
basalt
sandstone
schist

A LITTLE CHEMISTRY

Before discussing any more about minerals, let's step back and remember that minerals are made of **elements**. Well, we really need to step back even further to know that elements are made of **atoms**. Atoms are the building blocks of matter; they form elements when only one kind of atom is present. When we sing that "minerals on earth are found as element or compound," we are saying that minerals can either be made of one element, also called a **mineral element** (such as gold), or a combination of two or more different elements, called a **compound**. The term *mineral* includes both mineral elements and compounds.

And now, a little more chemistry. Elements are assigned particular **symbols**, such as Si for silicon, H for hydrogen, and Au for gold. When elements combine in compounds they can be written as a **chemical formula** which shows how many atoms of each element are in the compound. For example, the chemical formula for the most common mineral compound, **quartz**, or silicon dioxide, is SiO_2,

Quartz crystals
Minerals are crystals of various sizes and forms.

This miner is looking for Au in the H_2O.

CHARACTERISTICS

Minerals have particular characteristics that define and distinguish them from all other substances. They are:

1. **inorganic**—made from non-living substances, not decayed plant or animal parts. But some organisms, such as clams and snails secrete a mineral called calcite to form their shells.
2. **formed by nature**—made naturally. Man-made versions can be manufactured in labs, but they are not minerals. To the untrained eye, synthetic "diamonds" may look like those mined in Africa, but they aren't minerals.
3. **solid**—as opposed to a liquid or a gas. Their shape cannot easily be changed unless it is subjected to extreme pressure or temperature.
4. **formed with a crystal or geometric pattern**—they have an orderly arrangement of atoms in a repeating three-dimensional crystal pattern.
5. **formed with a constant chemical composition**—they combine in a specific way, with a specific chemical formula, as explained on the previous page.

Mining for borax in Death Valley, California, with a 20-mule team.

MINERAL CLASSIFICATION

To classify minerals, mineralogists use the following eight physical properties.

1. **Crystal form**, or **habit**, is the geometric shape or pattern in which the individual crystals of a mineral grow. Some forms are cubic, such as pyrite; others are needle-shaped, which show that crystals grew quickly in only one direction.
2. **Hardness** is how easily the mineral can be scratched. The scale in the box on the right was developed in the early 1800s by mineralogist Friedrich Mohs. A diamond, #10, is the hardest substance and can scratch all other minerals. Talc, #1, the softest mineral, can be scratched by any other.
3. **Color** is one way to recognize only a few minerals, such as gemstones.
4. **Luster** is the way a mineral reflects light. A

Hmmm... I wonder if this talc engagement ring will last? Talc is the softest mineral—it is even used for making baby powder!

Mohs scale of hardness:
Softest to hardest
1-talc
2-gypsum
3-calcite
4-fluorite
5-apatite
6-orthoclase
7-quartz
8-topaz
9-corundum
10-diamond

> Examples of luster
> Metallic:
>   pyrite
>   copper
> Nonmetallic:
>   quartz
>   talc
>   asbestos
>   gypsum

metallic luster is shiny—the mineral reflects light. Conversely, a mineral is considered **nonmetallic** if it is dull and doesn't reflect.

5-Cleavage is how it breaks along a plane of weakness to form flat surfaces.

6-Fracture is the way it breaks on surfaces other than the cleavage.

7-Streak is the color the powdered form of a mineral makes. To make a streak, a mineral is scratched against a rough surface, such as a chalkboard or porcelain tile.

8-Density, or **heft** (the heaviness of something being lifted), is determined by a mathematical formula—mass, or weight of the mineral, divided by volume. Geologists have also developed a variation of this formula called **specific gravity**—the weight of a mineral in air divided by the weight of an equal volume of water.[8]

CLASSIFYING MINERAL COMPOUNDS

Most minerals are compounds, combinations of elements, and fit into eight classes including:

1-silicates made of silicon, the major ingredient in the surface crust, combined with oxygen and sometimes other elements.
   Examples: quartz, feldspar

2-carbonates made of carbon, oxygen, and sometimes other elements.
   Example: calcite = $CaCO_3$

3-sulfates and **sulfides** made of sulfur, oxygen, and metallic elements.
   Example: pyrite = FeS_2

4-phosphates made of phosphorous and oxygen.
   Example: apatite

5-oxides made of oxygen usually combined with a metal, often result in ore from which metals can be mined.
   Example: hematite (mined for iron)

"Hey, Mom, are minerals in the earth the same kind of thing as what's in these vitamin and mineral pills?"

"Yes! But most minerals are not the kind that form rocks!"

MINERAL GROUPINGS

Gemstones, those minerals prized for their luster, quality, and color have been worn through the ages as adornment. Descriptive categories of gemstones include:

1-transparent gems—can see light through them (such as diamond, ruby, and emerald).

2-opaque gems—can't see through them (such as turquoise, jade, and malachite).

INDUSTRIAL USES

Metallic minerals, or **industrial-type metals**, are made of metal elements that share several characteristics—they have luster, are good conductors of heat, and are opaque and malleable. The list below shows only a few metallic minerals and their industrial uses.

Metal element	Mineral	Product
silver	silver	photographic material
copper	copper	electrical wire
lead	galena	TV tubes
iron	hematite	steel
chromium	chromite	stainless steel
tin	cassiterite	cans
aluminum	bauxite	aircraft
mercury	cinnabar	batteries

Industrial Minerals are used by industry to make everything from paint to food additives and building supplies. The minerals and products listed on this page only begin to illustrate their widespread and diverse use.

Mineral	Product
calcium carbonate	linoleum
talc	lipstick
vermiculite	soil amendment
gypsum	sheetrock
titanium	paint
graphite	pencils
bentonite	kitty litter

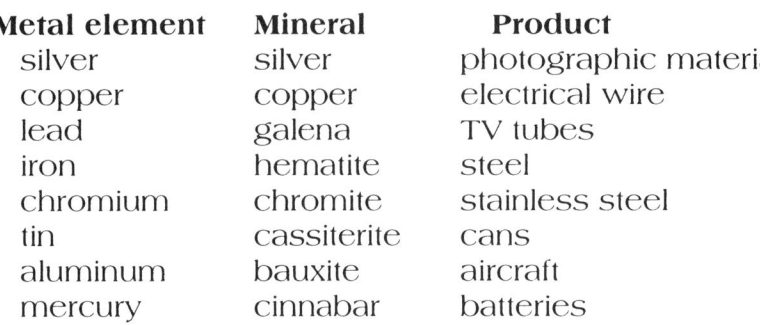

More industrial uses of minerals:
 carpets
 electric lights
 coffee pots
 linoleum
 fertilizers
 food additives
 lipstick
 cosmetics
 toothpaste
 powder
 sink cleaners
 cleaners
 kitty litter
 roofing material
 fiberglass
 tires
 mag wheels
 paint
 electronics
 computer chips
 gasoline
 lubricants
 concrete
 bricks
 sheetrock
 plaster of paris
 pencils
 carbon paper
 glossy paper
 inks
 pots
 pans
 metal castings
 plastics
 glass
 ceramics
 detergents
 antacids
 iodine
 paper[9]

During the course of the average American's life, he or she will use:
 3,600 lbs. of aluminum
 730 lbs. of zinc
 27,000 lbs. of clay
 35,000 lbs. of iron
 800 lbs. of lead
 1,500 lbs. of copper
 25,000 lbs. of salt[10]

Minerals are used in fireworks because they burn quickly and flash spectacular colors:

mineral	color
barium	bright green
strontium	deep reds
copper	blue
sodium	yellow
iron	gold
strontium and sodium	orange
titanium, zirconium, magnesium	silvery white
copper and strontium	lavender
aluminum	bright flashes loud bangs[11]

Some high-quality bicycles are made with combinations of molybdenum— often called "moly" (pronounced molly)— such as manganese and moly, and chrome and moly.

And now a few words about what minerals make
Rocks and the Rock Cycle

TYPES OF ROCK
Rocks are made up of minerals. Geologists classify rocks into three categories.

1. **Igneous rocks** are made of different kinds of magma that have hardened and the minerals in them have formed crystals. Igneous rocks are usually what makes up the other two kinds of rocks.
2. **Sedimentary rocks** are made of rock particles that have been compacted and cemented together, often in layers.
3. **Metamorphic rocks** are made of igneous or sedimentary rocks that have been changed by pressure and heat.

> **Bedrock** is the solid rock underneath soil and loose rock particles. An **outcrop** is bedrock visible on the surface.

Rockhound and his dog

ROCK CYCLE
The sequence of events in which rocks change into other kinds of rocks is called the **rock cycle**. Magma forms **igneous rock** as it cools underground or from a volcanic eruption. Water and wind break the rock down into small pieces and carry them into a low-lying area, such as a streambed or valley. As more sediment is deposited on top, the layers cement together to form **sedimentary rock**. Igneous and sedimentary rock can be baked and compressed into **metamorphic rock** by the heat and pressure of nearby magma or the movement of tectonic plates. Metamorphic rock and sedimentary rock can then become magma if they are heated to the point of melting, a process that happens deep in the earth such as at subduction zones. The cycle repeats itself when magma rises to the surface and cools to become igneous rock.

> What's the difference between rocks and minerals? Rocks are made of minerals—minerals are made of elements.

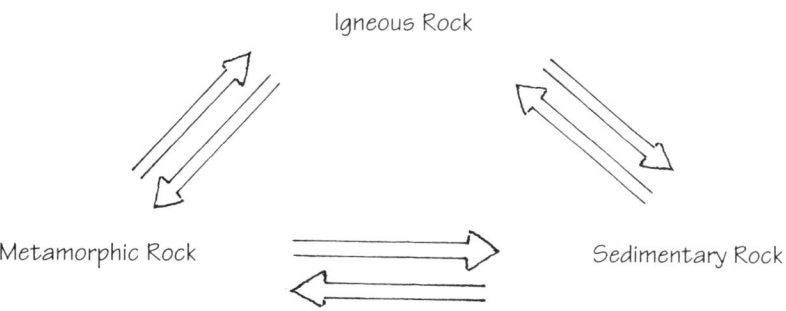

IGNEOUS ROCK

(to the tune of "Anchors Away")
"Anchors Away" is the official song of the U.S. Navy.

Igneous rock can be one of two kinds
Intrusive starts as magma and cools slowly underneath the
ground—like in granite, large crystals explain
Why this abundant rock has coarser texture, it's the coarser grain

Igneous can also be extrusive rock
When magma becomes lava at the surface from volcanoes
Where it is quickly cooled, crystals are small
Finer grain the finer texture—the most common is basalt

Magma can turn to glass when it cools fast
No crystals in the dark obsidian, pumice bubbles within
So for texture there are these three kinds
With crystals large, or crystals small, or with crystals not at all.

IGNEOUS ROCK

Previous chapters contain information about the kinds of landforms and geologic structures magma makes, including volcanoes, batholiths, and dikes. This chapter is all about magma on the smaller scale—the type of rocks it creates. "Igneous" comes from the Latin *ignis*, which means "fire," (as in ignite) and that's exactly what igneous rocks are formed from: fiery-hot magma that comes from deep within the earth.

TYPES OF IGNEOUS ROCK

Igneous rock is created as magma cools and hardens. During the period of cooling, minerals in the magma form crystals. This process takes place either where magma is trapped within the crust, or after it is spewed onto the earth's crust such as from a volcanic eruption. These two different locations affect the speed of cooling and, therefore, the size of the crystals, a key feature that distinguishes the two distinct groups of igneous rock.

> 1-**Intrusive igneous rocks**, also called **plutonic rocks**, are made from hot magma which forms from intrusions in the crust (see page 23). It cools slowly because the crust insulates it. Slow cooling allows time for minerals to grow into the large crystals characteristic of intrusive igneous rocks, such as **granite**, the most common.
>
> 2-**Extrusive igneous rocks** are created from volcanic eruptions which extrude (force out) mafic, intermediate, and felsic lavas onto the crust. There is no insulation as they hit the cooler air, causing quick cooling which does not allow time for crystals to grow. The results are tiny crystals in extrusive igneous rocks, such as **basalt**, the most common.

> The two places of igneous rock formation: in the crust (intrusive) and on the crust (extrusive).

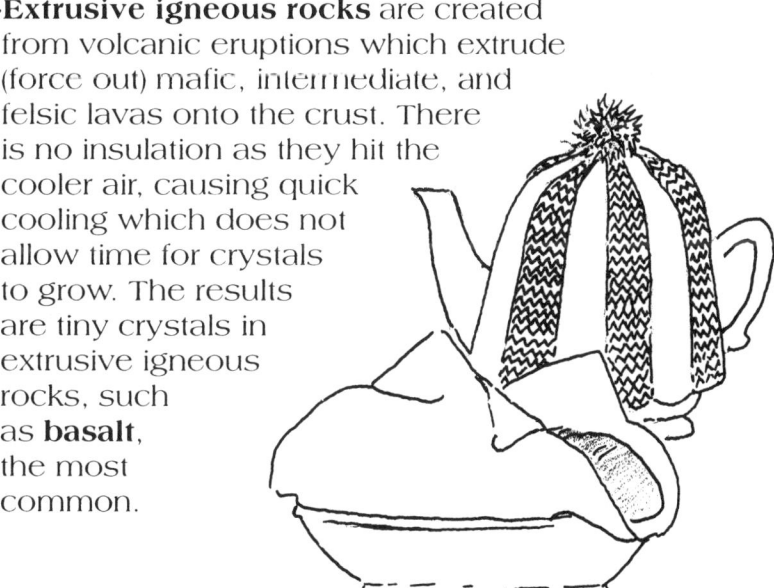

Like tea in a cozy or bread surrounded by a warm cloth, intrusive igneous rock was insulated deep in the earth's crust. Cooling was slow, so crystals had time to grow before they completely cooled and got hard.

CLASSIFICATION

Geologists classify igneous rocks according to the following two main characteristics:

1. **Texture**, or **grain**, is determined by the size of crystals. Rocks may have a rough **coarse grain**, a smoother **fine grain**, or a **glassy** texture.
2. **Chemical composition** is determined by the amount of silicates (see page 30) which make up most of the minerals in igneous rocks.

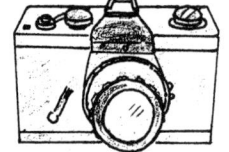

Crystals need time to grow, so as a result:
Slow cooling creates large crystals
Quick cooling creates small crystals

(Some things just need to be photographed instead of drawn!)

Texture

The first major way of classifying igneous rocks was developed in the 1800s and was based on texture. This was because the crystal size can be easily observed. Intrusive rocks have a **coarse grain**, large crystals easily seen with the naked eye. For example, the crystals in granite (mica, feldspar, and quartz) can all be seen without magnification.

Distinct minerals can be seen in this piece of granite.
1-mica, the darkest areas,
2-quartz, the slightly lighter,
3-feldspar, the lightest.

Small crystals form the **fine grain** of quickly cooled extrusive rocks. These crystals can only be seen under a microscope. Basalt has a fine grain.

Rocks with a **glassy** texture, called **volcanic glass**, cooled so rapidly that crystals couldn't grow. Black obsidian is dense and solid, while lighter-colored pumice is like a glass sponge because it is full of tiny pockets formed from trapped gasses.

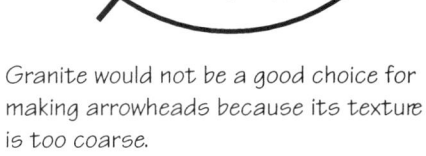

Granite would not be a good choice for making arrowheads because its texture is too coarse.

Arrowheads are often made of obsidian—its glassy texture gives them sharp edges—great for hunting!

The following chart summarizes how texture, place, and speed of cooling are related.

*Texture depends on the **place of cooling** which controls the **speed of cooling**.*

texture	place of cooling	speed of cooling	kind of rock	examples
coarse	in the crust	slow	intrusive	granite, gabbro
fine	on the crust	quick	extrusive	rhyolite, basalt, andesite
glassy	on the crust	very quick	extrusive	obsidian, pumice

CHEMICAL COMPOSITION

Classification based on chemical composition was developed later than that based on texture because more complex scientific instruments were needed to study igneous rock content. The chemical classification is the same as the chemical classification of magma (discussed in Chapter 4, "Volcanoes") because igneous rock is hardened magma. Classification in both cases depends on the amount of silicates.

    1-mafic—few silicates
    2-felsic—lots of silicates
    3-intermediate—intermediate amount of silicates.

As geologists studied the chemicals in igneous rocks, they discovered that rocks with the same composition can be either coarse or fine textured. Texture continued to depend on the location of the rock during its formation. The following chart shows how rocks of similar composition can, nevertheless, differ in texture.

Texture still depends on speed of cooling which is determined by the place of cooling.

lava composition	rock name	kind of rock	place of cooling	crystal size	texture
mafic	basalt	extrusive	on the crust	small	fine
	gabbro	intrusive	in the crust	large	coarse
felsic	rhyolite	extrusive	on the crust	small	fine
	granite	intrusive	in the crust	large	coarse
intermediate	andesite	extrusive	on the crust	small	fine
	diorite	intrusive	in the crust	large	coarse

SEDIMENTARY ROCK

(to the tune of "Bonny Blue Flag")

The melody comes from a little known-tune called "Irish Jaunting Car" about a ride on the railroad. But with other words written by Harry McCarthy in 1861, "Bonnie Blue Flag" became the Southern soldiers' favorite during the Civil War.

Rocks are broken down and as sediment carried away as
boulders, cobbles, pebbles, gravel, sand or silt or clay.
When they are deposited in layers they may all be cemented in conglomerate, sandstone or shale.

Chorus:
Rocks, rocks, sedimentary rocks:
Clastic, chemical, organic sedimentary rocks.

Minerals that were dissolved in water at one time
Precipitate or from evaporation are left behind
Rocks are formed from some of these chemical sediments:
Gypsum, rock salt, compact limestone, chalk, and chert or flint
Chorus

Organic limestone comes from hard remains of animals
Peat from plants decayed compresses to lignite then to coal
All three kinds of sedimentary rocks are valuable
In building and construction and in things industrial
Chorus

SEDIMENTARY ROCK

After igneous rock comes to the crust's surface from a volcanic eruption, wind and water eventually wear it down into broken, loose particles. These pieces, and pieces of other types of rock, are called **sediment** when they are carried away and laid down elsewhere. This sediment may eventually become **sedimentary rocks**, which, though they make up only 5 percent of the total crust, they cover 75 percent of its surface.

> Sedimentary rock covers more of the crust than any other kind of rock.

> "Sedimentary" comes from the Latin word *sedimentum*, which means "that which has settled."[12]

BEDDING

Sediment continues to move downhill and collects in a **deposition site**, a low-lying area such as a riverbed or valley, where it forms horizontal layers called **strata**, or **beds**. (The law of original horizontality is discussed on page 10.) These places of **sedimentation** show **stratification**, or layering, as new sediment settles on top. Sedimentary rocks are characterized by layering, or bedding, which can be of several types:

1. **Graded bedding**, the coarsest textured sediment is on the bottom and the finest on the top.
2. **Cross-bedding**, the sediment is deposited at different angles or slopes. A change in river flow or wind direction while sediment is laid down causes the angle changes seen in hardened sedimentary rock.
3. **Ripples**, the sediment forms beds of ridges that lie perpendicular to the water current (as in a river) or wind direction that deposits them. Ripples that become hardened sediment look just like the ripples of loose particles in shallow streams or sand dunes.

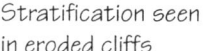
Stratification seen in eroded cliffs

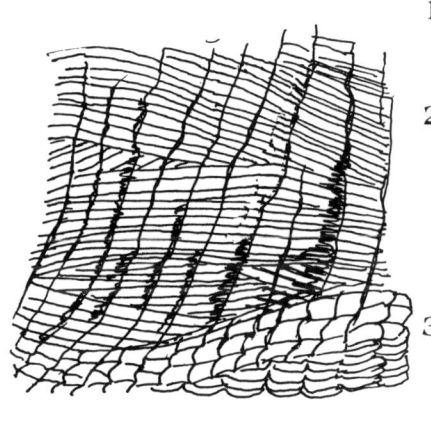
cross-bedding

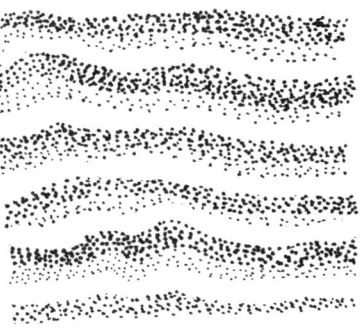
Ripples

The pressure from tons of sediment on top of the lower layers creates new rocks below. These sedimentary rocks are formed by a process called **lithification** which consists of **compaction**, the squeezing out of air and water; and **cementation**, the gluing of particles together into solid rock.

KINDS OF SEDIMENTARY ROCK
The three major types of sedimentary rock formed by these processes are:
1. **Clastic**—made of rock particles.
2. **Chemical**—formed from minerals once dissolved in water.
3. **Organic**—made of hard remains of animals, such as shells or corals.

Clastic sedimentary rocks (from the Greek *klastos* meaning "broken") are made of worn down, broken pieces of rock, or **clasts**. They may be the size of gravel, or smaller rock particles such as sand, silt, and the smallest, clay. Clastic rocks are classified by texture, which is related to the size of their particles.

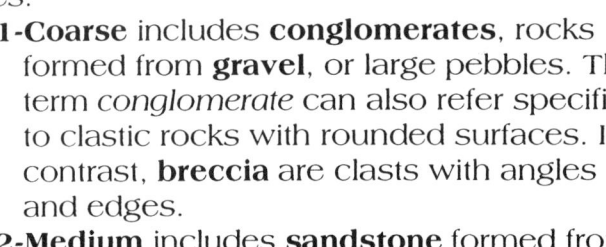

The rocks in breccia are angular.

1. **Coarse** includes **conglomerates**, rocks formed from **gravel**, or large pebbles. The term *conglomerate* can also refer specifically to clastic rocks with rounded surfaces. In contrast, **breccia** are clasts with angles and edges.
2. **Medium** includes **sandstone** formed from sand cemented and hardened into stone. It has been used through the centuries for making monuments and buildings—from the ancient Sphinx to the presidential White House.
3. **Fine** includes the following:
   siltstone, formed from silt;
   mudstone, made of mud, which is finer than silt;
   shale, the most common sedimentary rock, is formed from silt and clay;
   claystone, the finest sedimentary rock, is made of clay.

The rocks in conglomerates are rounded.

The Great Sphinx in Egypt is carved of sandstone.

Sedimentary Rock

Chemical sedimentary rocks are made of minerals that were once dissolved in water, usually seawater. Two main types are created.

1. **Precipitates** are formed when minerals fall out of the water (precipitate)—like rain falling out of the sky. Precipitates are laid down underwater and harden.
   Examples include:
   - a-**compact limestone**, the most common precipitate, made of the mineral calcite from seawater.
   - b-**chert**, also called **flint**, a rock most commonly made of microscopic silica shells.
2. **Evaporites** are what remain after water has evaporated, or become vapor.
   Examples include:
   - a-**gypsum rock**, a common evaporite of seawater formed into thick beds and commonly used in sheetrock for walls inside buildings. Gypsum rock is composed mainly of the mineral gypsum.
   - b-**rock salt**, another common evaporite of seawater, formed mainly from halite and used in foods.

Chalk is made from limestone.

The flintlock was used during the Revolutionary War. When struck by the hammer, the flint on the plate produced a spark that ignited the priming powder and caused the gun to fire.

Organic sedimentary rocks are defined as those formed from living, or **organic**, materials, such as the remains of animals or plants. Compressed plant and animal parts can become rocks!

1. **Organic limestone** is made of corals and shells.
2. **Coals** are formed from buried plant materials and changed over long periods of time by pressure from the weight of overlying layers. Pressure increases with burial depth. **Peat**, composed of 75 percent water, has a shallow burial in a swamp and bog but is the first stage in the development of hard coal. As burial depth increases, **lignite** is the next stage of coal. Plants are then pressured into **brown coal**, then **bituminous coal**, and finally, into the hardest type, **anthracite**.

"I wish I could be a rock!" Will this fern get its wish? Well, yes, with time and pressure it can be changed into peat, lignite, or coal.

Gas and oil, obviously, can't be considered sedimentary rocks because of their vaporous and liquid states.—Nevertheless, like coal, they are **fossils fuels** formed from sedimentary processes and found in sedimentary rocks.

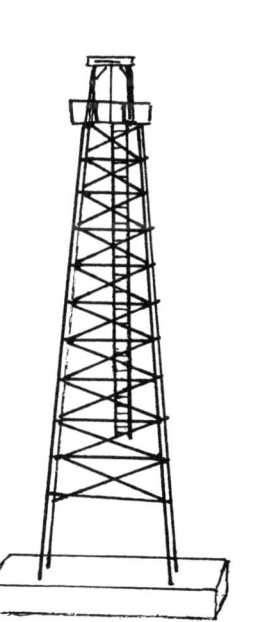

Oil derrick

METAMORPHIC ROCK

(to the tune of "Just Before the Battle, Mother")
Written by George Root, the most prolific song writer during the Civil War. The song was so moving and sad, some officers did not allow their soldiers to sing it.

The kind of rock called me-ta-mor-phic, is made of rocks that have been changed. Perhaps in re-crys-ta-li-za-tion (a change in how they are ar-ranged); or in size or shape of cry-stals, affec-ting tex-ture and the grain; or change in min-eral com-po-si-tion— which means the rock that was is not the same.

If you ask what cause these chang-es mak-ing rock so hard and dense. I'll tell you it's from heat and pres-sure from the mov-ing con-ti-nents heat from in-trud-ing near-by mag-ma in moun-tain buil-ding or the weight, of tons of sed-i-menta-ry stra-ta or faulting fold-ing from tec-ton-ic plates.

Foliation looks like layers
Shale or siltstone becomes slate,
With more heat and with more pressure
Hardens to phyllite, or schist, or gneiss,
The metamorphic rock that shows no layers
Nonfoliated rock has layers not seen:
Limestone hardens into marble,
And basalt to serpentine

METAMORPHIC ROCK

Metamorphic rock is rock that has been changed from one kind into another. "Metamorphosis" is from the Greek words *meta* and *morphe*, meaning "to change in form." That is exactly what happens: igneous or sedimentary rocks change their texture and mineral composition—to become more compact and dense metamorphic rocks. The minerals **recrystallize**, or change into new grains of crystals which are usually larger than the previous ones. These changes are produced by:

    1-**pressure** as overlying layers squeeze the rocks below, or pressure from tectonic forces as they squeeze and deform the crust.

    2-**heat** from nearby magma that's hot enough to bake the rock, but not to melt it.

TYPES OF METAMORPHISM

A combination of heat, pressure, and even water produces five types of metamorphism, but the two most common types are:

    1-**regional**
    2-**contact**

These metamorphic processes produce rock with different textures and appearances.

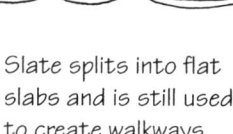

Regional metamorphism relies mainly on pressure from tectonic plates folding and faulting during mountain building. The **parent rock**, or original rock, may be sedimentary, igneous, or even metamorphic rock that becomes buried in the earth, then is squeezed by the pressure of tectonic forces. As the rock is pushed down deeper, the temperature rises and the heat necessary for metamorphosis is supplied by heat inside the earth.

Slate splits into flat slabs and is still used to create walkways.

Slate for chalkboards

Regional metamorphism creates rocks with textures that are **foliated**, which is the lining up of minerals on a plane such as seen in **slate** when it is split into flat slabs. Slate has long been used for roofing and walkways, and was once used for chalkboards in classrooms of the 1800s.

Banding, the striped appearance seen in **gneiss**, is caused by the same kind of minerals banding together and recrystallizing as the new rock is formed.

> Foliation and banding are not layering, the sedimentary process creating layers of sediment in beds.

CLASSIFICATION

More heat and more pressure cause further metamorphosis. This means that the same parent rock can metamorphose into **low-grade** (with the least amount of metamorphosis) rock such as slate, and also **high-grade** (with the most metamorphosis) rock such as gneiss. Foliated metamorphic rocks are classified by their degree of metamorphosis and amount of foliation. Some common foliated rocks, all formed from the parent sedimentary rock shale, have the following classifications:

1. **slate**—lowest grade. Foliated in close, thin bands seen only with a microscope.
2. **phyllite**—slightly higher grade of metamorphism. Phyllite is glossy because its mica crystals have had more time to grow and get larger. It has microscopic foliation like slate, but it tends to have a wavy or rippled appearance.
3. **schist**—moderate grade. The foliation is coarse, which means micas and other minerals can be more easily seen. Schist is one of the most common metamorphic rocks.
4. **gneiss** (pronounced "nice")—high grade. The foliation is coarser with wider light and dark bands of minerals.

With increases in pressure and heat, the parent rock metamorphoses into different forms of new rock.

parent rock
shale

slate

phyllite

schist

gneiss

Contact metamorphism relies less on pressure and more on the heat of nearby magma. Country rock, the rock at an igneous intrusion such as a dike or still, is baked by the heat. This process produces rocks with textures that are **nonfoliated** or **granular**, and do not have a banded appearance. Examples include:

1. **marble**—formed from the mineral calcite and sedimentary rock, either limestone or dolomite.
2. **quartzite**—formed from the mineral quartz and the sedimentary rock, sandstone.
3. **serpentinite**—formed from the mineral serpentine and the igneous rock, basalt.

Statues such as Michelangelo's "David" are often carved in marble

In addition to the United States Capitol building, many government buildings in Washington D.C., such as the Supreme Court Building, the Washington Monument, the Lincoln Memorial, and the Jefferson Memorial, are made of white marble. Marble is symbolic and stands for democracy. American political buildings imitate ancient temples of Greece—the birth place of democracy.

United States Capitol

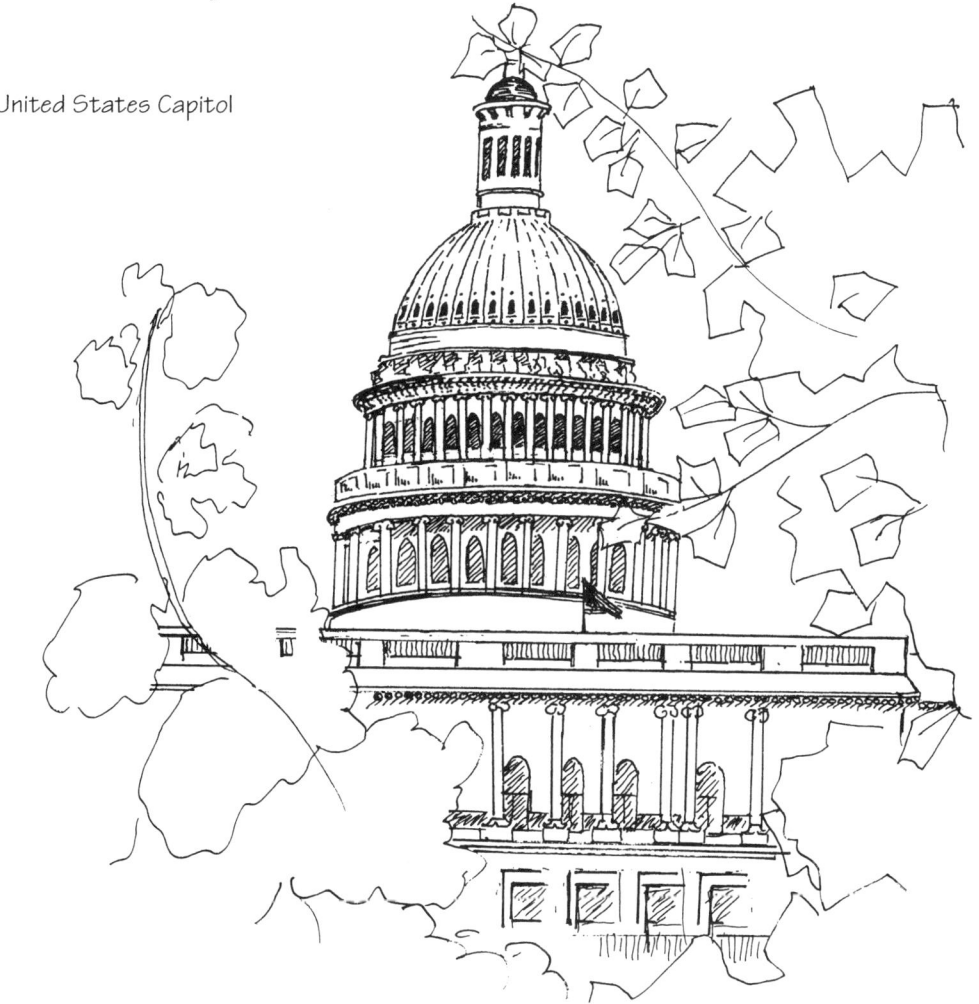

Summary of facts about several metamorphic rocks.

metamorphic rock	typical parent rock	parent rock type	texture	grade
slate	shale	sedimentary	foliated	low
phyllite	shale	sedimentary	foliated	low
schist	shale	sedimentary	foliated	moderate
gneiss	shale	sedimentary	foliated	high
marble	limestone	sedimentary	nonfoliated	variable
quartzite	sandstone	sedimentary	nonfoliated	variable
serpentinite	basalt	igneous	nonfoliated	variable
argillite	mudstone	sedimentary	nonfoliated	low
greenstones	basalt, ash	igneous	nonfoliated	low
amphibolite	basalt (feldspar)	igneous	nonfoliated	high

WEATHERING OF ROCKS

(to the tune of "Wearing of the Green")
This early Irish melody was popularized with lyrics written during the
1840s lamenting England's harsh domination over Ireland:
"St. Patrick's Day no more we'll keep, his color can't be seen.
For there's a bloody law again' (st) the wearing of the green."

In the changing of the landscape a most important thing is
how the rocks are broken down: we call it weathering. There are two kinds of weathering:
one is physical; decomposition is the other, known as chemical. And four
things that affect the rate of all weathering: composition of the rock; the climate
(temperature and rain); And topography's exposure to the wind and rain and sun; but
vegetation may provide the rock some protection.

There are four kinds of weathering we say are physical:
Disintegration happens, and it is mechanical:
Exfoliation is the flaking off of rocks in sheets
When rock expands resulting in surface cracks and joints
Another is frost action: freezing water expands cracks
And from growing plants the prying roots do slow, destructive acts
And finally there's abrasion from physical contact
When edges sharp are rounded, by both grinding and impact

Now to all those kinds of weathering with chemical compare
Which decomposes rock by oxidation from the air;
Or hydrolysis from water; or by acid from the rain;
Or from the lichen growing; or from plants decayin'

WEATHERING OF ROCKS

Two processes are constantly at work changing the surface crust of the earth: construction, from the movement of tectonic plates and volcanic activity; and destruction, from the **weathering** away or breaking down of the crust. Without either process we'd be in trouble. Without construction the earth would become as smooth as a marble—with no mountains. Without destruction, it would be uninhabitable—no soil, no plains, no valleys—we'd all be living at very high elevations!

Physical weathering

It may seem surprising that something as hard as a rock could be broken and worn down into smaller particles, but it happens— through the weathering process. Little by little, grain by grain, weathering wears down the earth's crust and produces what the rivers carry away and deposit as sediment to make sedimentary rocks. Clays, silt, and sand—and the soil they create—are all products of weathering. Two distinctive types of weathering break down hard rock.

1-**Physical weathering**, called **disintegration**, is a mechanical process that breaks them down.

2-**Chemical weathering**, called **decomposition**, is a chemical breaking down of rocks caused by acids in the soil, air, or water.

CONDITIONS AFFECTING RATE OF WEATHERING
The amount and speed of weathering (both physical and chemical) are affected by several factors.

1-**Type of rock**—a soft rock will break down before a hard rock. For example: intrusive rock, such as granite, is much harder than sandstone. As a result, intrusive igneous rock formations, such as those discussed on page 23, remain hidden until outer, softer rock has been weathered away.

Chemical weathering

2-**Climate**—water increases the speed of weathering. Rocks in areas receiving little rainfall, such as deserts, weather slower than those in moist mountain forests. Also, climates with extreme day-to-night or seasonal temperature changes cause rocks to weather quickly due to freezing and thawing. In contrast, mild temperate climates with

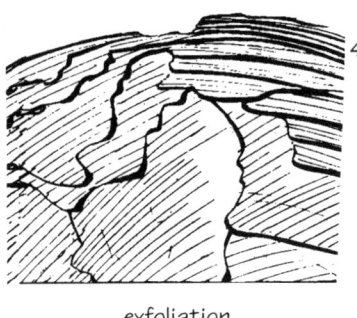
exfoliation

little temperature change, weather rocks slowly.

3-Topography—the shape of the land affects weathering. Mountain tops are more exposed than low-lying valleys, causing rock on them to weather more quickly.

4-Vegetation—plants covering rocks can decrease the speed of physical weathering because they shield them from wind, rain, and the heat of the sun. But plants may also increase chemical weathering because they, and the soils they grow in, may contain bacteria and acids which eat away at the rocks.

PHYSICAL WEATHERING

Disintegration is the physical process of rocks breaking into small particles. The four ways it is accomplished are listed.

1. **Exfoliation** is the flaking off of layers or sheets of rock, often from the release of pressure when overlying materials are weathered away. This causes the rock to expand slightly and creates cracks in the surface.

2. **Frost action**, or frost wedging, is caused by water getting into cracks and expanding as it freezes, making the cracks larger (like a sealed container that is too full of liquid pops and cracks when frozen).

3. **Root pry** results when a root grows into a small crack, or fracture, in a rock. As the root continues to grow, the crack expands, which increases pressure on the entire rock and creates even more fracturing.

root pry

4. **Abrasion** results from rocks tumbling and bouncing against one another, such as in a stream or river. The rocks become smaller as rough edges are worn down, creating smooth and rounded river stones and pebbles.

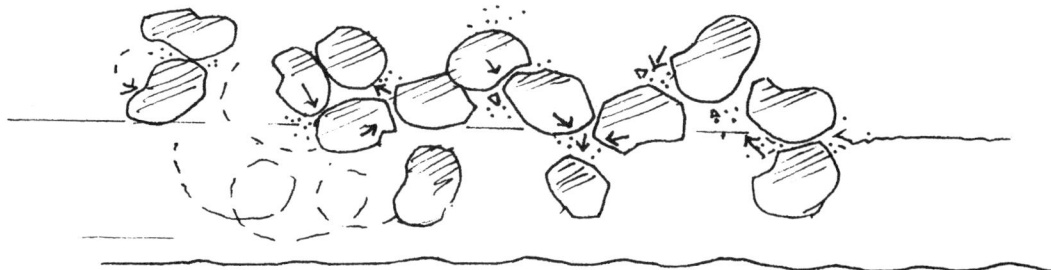

Rocks tumbling together in a stream of river cause them to become rounded and smoothed from abrasion.

Weathering of Rocks

CHEMICAL WEATHERING

Rocks can be broken down by their interaction with chemicals in the air and water in a process called **decomposition**. In this kind of weathering, rocks slowly dissolve as their chemical compositions are changed. Four kinds of chemical weathering include:

1. **Oxidation**, when **oxygen** in air combines with minerals in rock to form oxides (see page 39), which usually cause a noticeable color change. Copper oxide creates a green, turquoise color; iron oxide (also called rust) creates, obviously, a rust color. Rust develops on rocks composed of iron, just as it does on old bikes and metal junk.

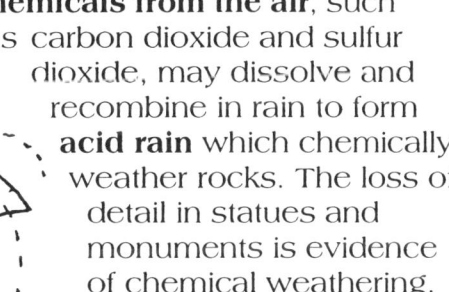

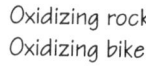
Oxidizing rock
Oxidizing bike

2. **Hydrolysis**, the process of **water** combining with minerals in rock to break it down and create new substances. When the mineral feldspar in igneous rock undergoes hydrolysis, it forms clay.

3. **Plants** and lichens that grow on rocks emit acids that cause weathering. Decomposing plant material in the soil also creates acids that chemically break down rocks.

4. **Chemicals from the air**, such as carbon dioxide and sulfur dioxide, may dissolve and recombine in rain to form **acid rain** which chemically weather rocks. The loss of detail in statues and monuments is evidence of chemical weathering.

Famous stone statues, monuments, and buildings are slowly decomposed by the acid in rain (acid rain), which may occur naturally, but is definitely increased by air pollutants. The Statue of Liberty needed large scale reconstruction because the copper (a mineral element) plates were badly damaged by acid rain.

HYDROLOGY AND EROSION

(to the tune of "Yankee Doodle Dandy")
This is only one of George M. Cohan's popular dandy tunes written during the early part of the 20th century.

Wa-ter mov-ing and ef-fects it has: the sci-ence of hy-dro-lo-gy. In-cludes the wa-ter cy-cle as it moves from air to the ground to the sea. From o-ceans much e-va-por-a-tion; clouds are con-den-sa-tion in the air; Rain and snow's pre-ci-pi-ta-tion; ground wa-ter's from in-fil-tra-tion; sur-face run-off flows to-wards the sea.

Volume is amount of water, velocity is speed as water flows
Load's amount of sediment carried; depending on volume and speed
Capacity is the ability of water to carry sediment
Erosion is the movement of the sediment by runoff
Deposition drops the load when water slows

The amount of runoff is determined partly by topography:
Steeper gradient or slope will mean, the faster the runoff will be
Rate of precipitation effects runoff, vegetation slows the water down.
Thicker humus and the looser soil will also
Make more water soak into the ground

HYDROLOGY AND EROSION

Water is abundant, covering over 75 percent of the earth's surface. It is mainly stored in oceans but also in other **reservoirs** (places of water storage), such as glaciers, lakes, rivers, under the ground, and even in the atmosphere. Water changes its state—its physical condition—from liquid (water) to gas (vapor) to solid (ice) and back again as it cycles throughout the earth from one reservoir to another. The scientific study of water in all its various states, places, and ways of flowing is called **hydrology**.

THE WATER CYCLE

The **water cycle**, or **hydrologic cycle**, is the process of water moving between reservoirs. Major components of this continuous cycle are:

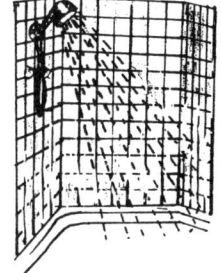

Condensation collects on steamy bathroom walls.

1. **evaporation**, the changing of water into vapor. This transformation happens wherever water is exposed to air; at the surface of oceans, lakes, man-made reservoirs, ponds, pools, and puddles. You can see evaporation when steam (vapor) rises from a hot kettle or cup of soup, or when the warm sun comes out after rain.
2. **condensation**, the changing of vapor into water; the formation of droplets. Moisture, as vapor in the air, becomes water and creates clouds as the rising air cools. You can also see condensation when warm air meets cooler walls and mirrors in a hot, steamy bathroom or on the outside of a glass of cold drink on a hot day.
3. **precipitation**, the falling of water onto the land as rain or snow. Droplets grow heavy enough to fall to the earth by the pull of gravity.
4. **infiltration**, the seeping of water into the ground through openings such as cracks or **pores**, the tiny spaces in or between rock particles.
5. **runoff**, the transporting of water in streams and rivers back to oceans.

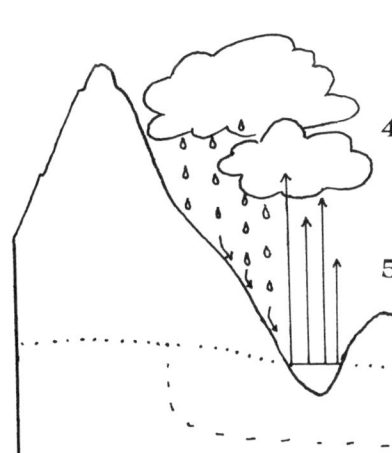
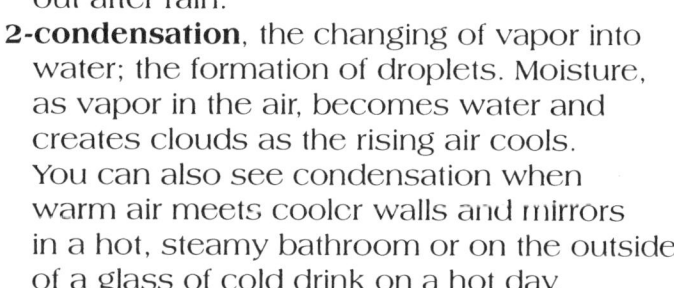

The water cycle

> CAUTION
> Two different terms that are often confused.
> **Weathering** is only the breaking down of rock particles—it does not move them.
> **Erosion** is the combined processes of picking up and moving weathered rock particles, and depositing them in another place.

RUNOFF AND EROSION

When all the pores in the ground are filled with water, it is **saturated**—no more water can soak in. Precipitation that does not soak in, or infiltrate, collects on the surface and becomes **runoff** that starts traveling downhill on a journey in streams and rivers toward the ocean. As runoff flows, it picks up and carries away weathered rock particles, called **sediment**, in the major geological process of **erosion** (discussed in detail on page 63). **Deposition**, the last step of this process, occurs later when runoff drops, or deposits, the sediment downstream. The differing sizes and weights of the rock particles determine how a river carries them, and when and where they are deposited. Sediment is transported as one of the following.

  1. **Dissolved load**; includes materials such as minerals and salts transported in **solution**. It makes up 20% of the total load and is the reason the oceans are salty and soil downriver is rich in minerals.
  2. **Suspended load**; includes small particles, such as sand, silt, and clay that are lifted up and held in **suspension** by the water as it flows. Suspended load is also called the **wash load**.
  3. **Bed load**; includes the larger rocks propelled along on the bottom, or **bed**, of a river by the force of the water.

*If all the pores are full of water it means the ground is **saturated** and the surface water becomes runoff, kind of like a sponge that is full of water and can't absorb another drop.*

A stream or river's total **load** is what is carried by the water. The load a stream can carry is dependent upon:

  1. **velocity**—the speed of the water. Fast water can carry a greater load than slow water.
  2. **volume**—the amount of water. **Discharge** is the volume of water passing a point in a given time. Streams and rivers with greater discharge can carry more load than those with lower discharge.

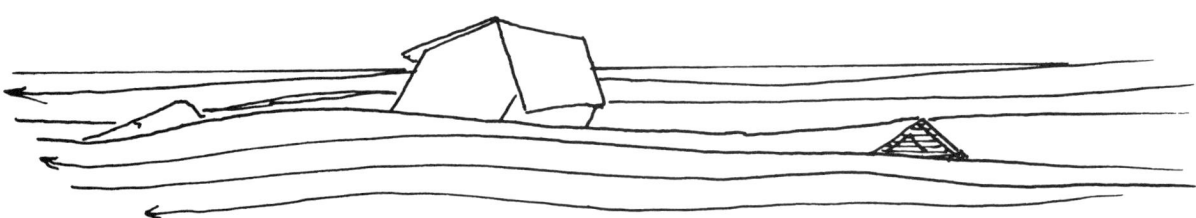

Floodwaters have tremendous velocity and volume. During a flood you may see a refrigerator, or even a house float by! They aren't the usual geologic load but they do illustrate important river concepts—a faster current such as a raging river (high velocity, volume, and discharge) can carry more and heavier objects than a slow stream (low velocity, volume, and discharge).

RATE OF RUNOFF

Several factors affect the rate of runoff—how much water there is, and how fast it moves.

1-Rate of precipitation—heavy rains create more runoff, even flooding, because the ground can't soak up water fast enough. There's just not enough time and space. The result of more runoff is more erosion—more rocks and particles moved and deposited downstream.

Result:
higher rate of precipitation = more runoff

2-Topography—steep slopes create faster runoff than gentle slopes or flat areas, with the same result as above—more runoff causes more erosion.

Result:
steeper slopes = more runoff

3-Vegetation—plants decrease the speed of runoff because they absorb water through their roots, create surface areas to which water adheres, and cause resistance to water's downhill movement. Vegetation is often planted on hillsides to slow down runoff and hold soil particles together, thereby controlling erosion.

Result:
more vegetation = less runoff

4-Other surface factors such as thick, loose soil and **humus** (organic matter in the soil), slow down runoff because they are very **porous** (full of pores or spaces), providing more places for water to soak in.

Result:
more soil and humus = less runoff

Vegetation soaks up water and holds soil and rock particles together.

DEPOSITION

Sediment stays suspended in streams while the water is turbulent, but as it slows down gravity causes the sediment to drop out. In simple terms: when a stream slows, it drops its load. This process is called **deposition**, the final step of erosion. The largest, heaviest rocks are the first to be deposited as they stop their downstream tumbling, next comes sand, followed by silt, and then finally clay, the smallest of the particles.

*Deposition from streams and rivers is well sorted or **stratified**, meaning that particles of the same size and weight are deposited together, similar to how the sediment in this jar has separated.*

WATERWAYS AND EROSION

(to the tune of "Shenandoah")

This is one of the most beautiful traditional tunes. There is speculation as to its origins but one source traces it to American and Canadian voyageurs; in which a white trader courts the daughter of Chief Shenandoah and carries her across the wide Missouri River.

Erosion is the process that moves
Rock and soil—the land reshaping
When water, wind, and glaciers carry
And wear away—
Erode away
Highlands into lowlands

An area drained by a river
And its branching tributaries—
A watershed is separated
By a divide
And on each side
Water flows in two directions

A river's source is the headwaters
Making channels and V-shaped valleys
Straight, steep, and wild; rapid erosion
Rivers flow
While they erode
The ground beneath them

Volumes rise, high deposition
Meanders on the flat wide floodplains
Continues to mouth or base level
Forms deltas there
And that is where
It flows into the ocean

WATERWAYS AND EROSION

PROCESSES OF EROSION

The most important agent in sculpting and shaping the earth is water. Running water demonstrates the three processes known collectively as erosion:

1-erosion—the picking up of weathered rock particles called sediment

2-transportation—the carrying away of sediment

3-deposition—the dropping of sediment downstream

Erosion is also accomplished by wind, gravity (mass movement), and ice (glaciers).

> The Nile River in Africa is the world's longest river, and runs 4,145 miles (6,632 km) from its source above Lake Victoria to its mouth at the Mediterranean Sea.

> The Amazon River in South America is the largest river, filling the ocean with 20 percent of the volume of all waters from rivers. It's the second longest river, 4,080 miles (6,528 km) long.

1-Erosion   2-Transportation   3-Deposition

Although this dump truck illustrates the three factors in the erosion process, it can never move as much sediment as rivers.

RIVERS

A stream picks up sediment near its **source**, or **headwaters**, and deposits it near its **mouth**, an opening to a large body of water such as a lake or ocean. Near its source, where it begins in the mountains, a stream or river is usually straight, steep, and wild. The sharp drop in elevation causes the river to flow quickly and take the shortest route possible. The river erodes its bed cutting deep gouges. It carves a V-shaped channel with steep-sided riverbanks, and may create **waterfalls** if the path is over a cliff of hard rock. **Rapids** are formed where the water is forced to navigate around other hard, more weather-resistant rock on steep slopes. As the stream continues its downhill course to the ocean, it picks up and carries sediment from the surrounding land.

> Just a few old and modern watercraft: canoes, barges, flatboats, riverboats, kayaks, rafts, canal boats, rubber rafts, drift boats.

Waterfalls are usually near the headwaters of a river.

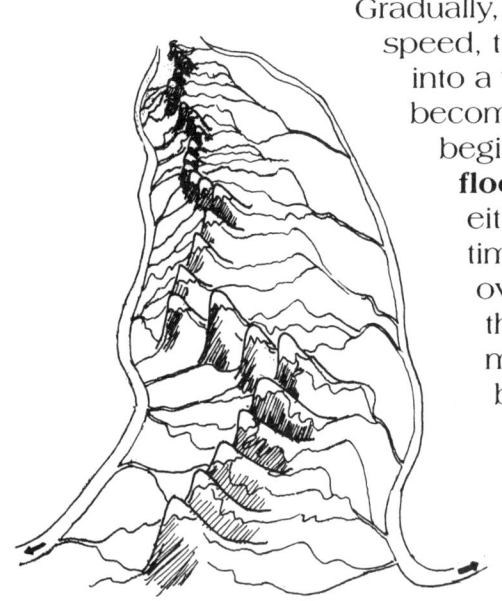

Gradually, as the river has less slope and speed, the V-shape of the channel widens into a valley and the steep riverbanks become gently sloped. The stream begins to drop its load, creating a **floodplain**, a large, flat area on either side of the riverbank. During times of flooding, the river will overflow its banks and spill onto the floodplain, bringing with it more sediment. The soil created by this deposition is good for growing crops, but those farming along the river's edge brave the possibility of floods in exchange for the valuable crops the fertile soil produces.

A **divide** is high ground, such as a mountain range, that separates streams or rivers flowing in opposite directions. The Continental Divide of the Rocky Mountains is a famous example.

FLOODPLAIN CHARACTERISTICS

On the floodplain, the water may form a **braided stream**, or **braided channel**, where the river divides and rejoins. The river often forms **meanders**, which are curves and loops across its floodplain. **Oxbow lakes** are created when these loops are cut off from the river. **Sandbars** are made when the river deposits its load right in its own bed, creating dangerous obstacles for river craft. As sand and gravel are repeatedly dropped at the tops of riverbanks, naturally formed **levees** (embankments) are created.

Floodplains are good places for farming because the ground is enriched by the deposition of the river's load when it overruns its banks.

The area draining into a particular river is called a **river basin, drainage basin,** or **watershed**.

When the river reaches the ocean, it creates a **delta** in the shape of a fan, as it finally deposits its lightest and smallest particles, such as silt and clay. The place where the river enters the ocean or a lake is its **mouth**, also called **base level** because the river cannot erode past this point.

In an early North American land claim, France claimed all the land whose waters drained into the Mississippi. That's all the land east of the Continental Divide in the Rocky Mountains and west of the crest of the Appalachians!

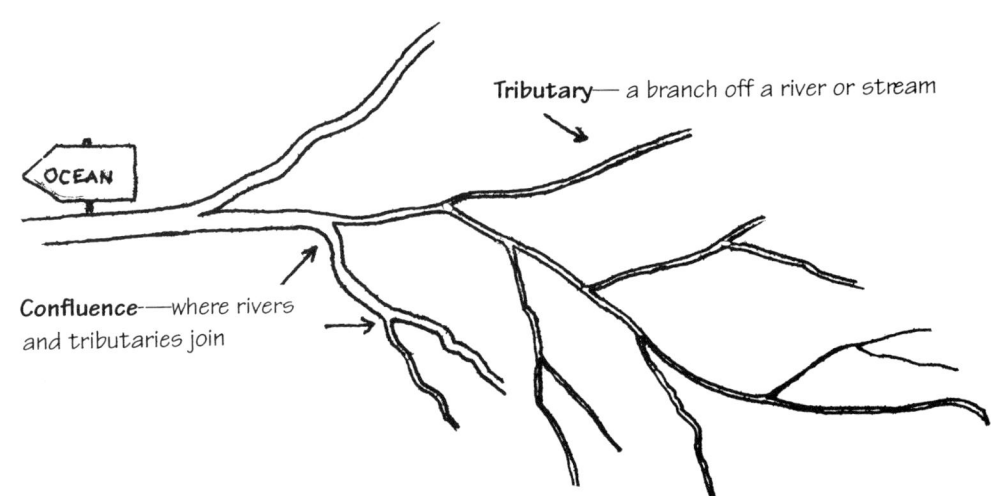

Tributary— a branch off a river or stream

Confluence—where rivers and tributaries join

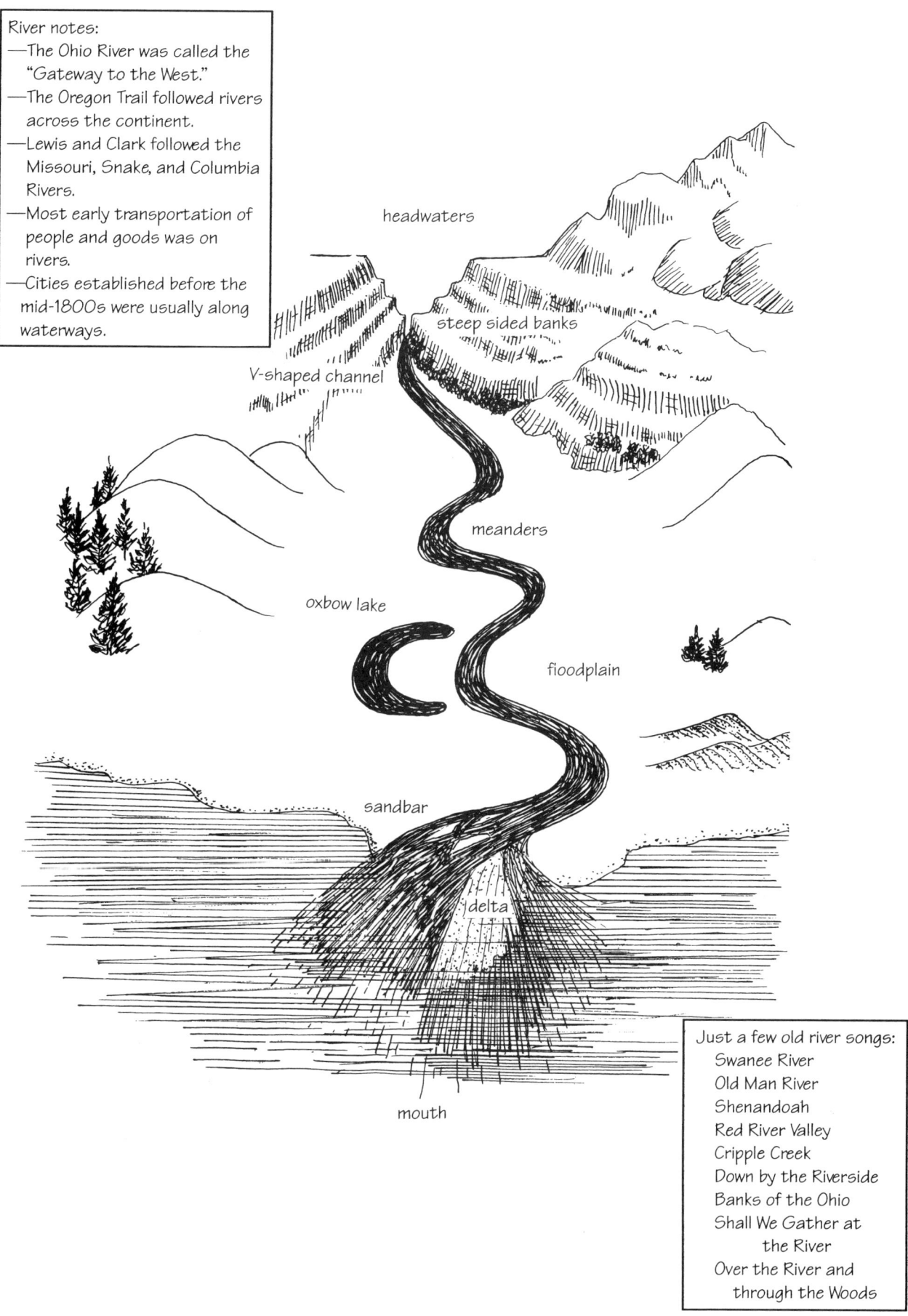

GROUNDWATER

(to the tune of Handel's "Water Music Suite")

George Frideric Handel wrote this piece in 1719 for boating on the Thames River in England.

Wa-ter un-der ground, flow-ing un-der ground, through rock that's por-ous and permeable.

Por-ous rock is like a sponge; and if the pores are con-nec-ted rock is per-meable.

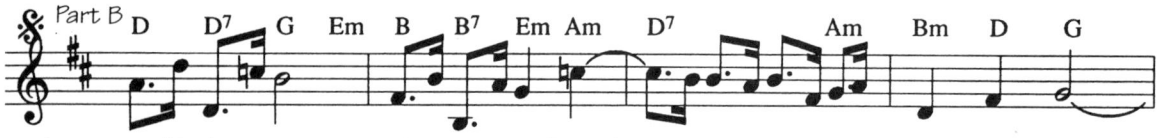
A permea-ble la-yer, is an a-qui-fer and a spring is ground wa-ter that comes to

the sur-face na-tur-ally, and wa-ter tables may be, at the sur-face in a mar-sh lake or stream. Ar-

tesion wells from pressure rise a-bove a-qui-fers. Wa-ter under ground, flow-ing underground, through

caves and ca-verns it e-rodes. Minerals from chemical weathering are dissolved and

may be de-po-si-ted.

Part A: Water passes through
Slowly continues
Seeping down through the aeration zone
To the water table,
   where the pores are full of water
   —saturation zone

Part B: On hanging stalactites
Onto stalagmites
And in a column they may join
These calcium carbonate
Formations may be great
All this from groundwater

Ending:
Water underground
Slowly flowing
Underground

GROUNDWATER

When rain falls to the ground, it either becomes runoff making a path to the ocean, or it becomes **groundwater** infiltrating, or sinking, into the ground. Water seeps into the pores, or spaces, between grains and particles of rock. There is a lot of water under ground —the amount of groundwater is more than 100 times the amount in rivers and lakes;[13] and 22 percent of all fresh water in the world is stored in the ground![14] Throughout history people have literally tapped into this valuable resource and relied on groundwater for their basic water supply. In fact, its overuse is becoming a problem for growing communities. More water is being pumped out of the ground than is infiltrating back in.

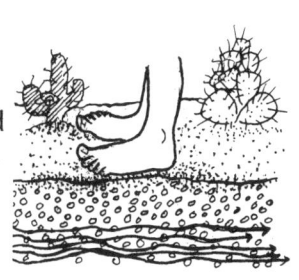

You may not be aware of it but there just might be water under your feet.

In order for water to penetrate the ground and flow through it, the ground must be:
- **1-porous**—have pores, tiny openings, enabling it to hold water.
- **2-permeable**—have pores that connect, creating passageways that water can flow through.

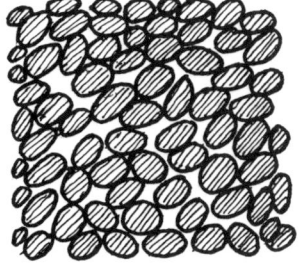

Porous ground

rock particles

pores

POROSITY AND PERMEABILITY

Porous rock or ground is like a sponge: it has spaces in it which can be filled with water or air. The amount of pores it has is called its **porosity**, and determines how much water the rock or ground can hold. Porosity is dependent upon how the particles are packed together. If rock particles are relatively loose, as in sedimentary rocks or a mixture of gravel and sand, there are lots of places to hold water, and the rock or ground is considered porous. Conversely, if particles are close and tight, as in many types of igneous and metamorphic rocks or clay soil, there are few pores and the rock is considered **nonporous**.

Rock or ground that is **permeable** allows water to flow into and through it because pores are connected, creating passageways between them. Good soil is permeable; but **concrete**, a mixture of water, sand, and rock particles that dries into a hard, dense building material, is **impermeable** —the pores and passageways are filled. Not only is concrete impermeable, but also nonporous.

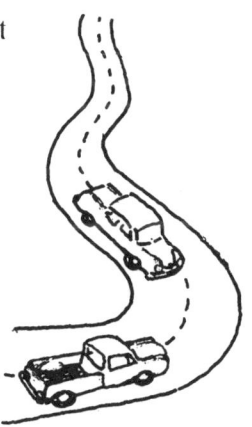

A modern problem: Pavement for roads and driveways creates impermeable layers that water cannot penetrate, which creates more runoff and more erosion.

INFILTRATION: THE DOWNWARD JOURNEY

Pores may be filled with water or air. If they are filled with air, or a mix of air and water, the ground is unfilled, or **unsaturated**, forming an area called an **aeration zone** or **unsaturated zone**. If water seeps into all the pores the ground will be full of water, creating a **saturated area**, or **saturation zone**. When farmers and homeowners dig wells, holes are drilled through the aeration zone, past the **water table** (the boundary between the two) and into the saturation zone.

The water table can be under the ground or on the surface, as the observable water level of a lake or stream. You can find an underground water table by digging a hole at the beach near the ocean's edge. The sand may be wet (some, but not all pores are holding water), but you have not reached the water table until water seeps into the hole and collects.

Precipitation makes the water table rise, but discharge (as water leaking out at a lower elevation) and pumping (mechanical removal of water) lowers it. A stream may flow under the ground during a dry spell and reappear on the surface in its streambed during a wet season.

Beneath the water table is the **aquifer**, the storage area for water amid gravel and other rock particles. Water travels horizontally through the aquifer by way of connecting pores of the permeable rock. Below it there may be types of impermeable igneous or metamorphic rock that the water cannot infiltrate so it can't flow downward any farther.

Wells are dug or drilled to get down to the aquifer and to pump groundwater up to the surface. In an **artesian well**, water rises to the surface naturally when pressure from water draining down from higher ground above the well forces a flow of water upward. A **spring** is where groundwater flows naturally from the ground onto the surface or into a body of water.

Wind powered pump to draw water from the ground

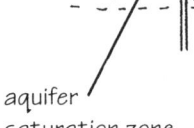

water table

aquifer
saturation zone

Artesian well demonstration

Artesian well: water draining from higher ground forces water at a lower level upwards to the surface.

Groundwater

GROUNDWATER EROSION

Some **caves** are created by groundwater dissolving, or mixing with rock, and carrying it away. Caves are commonly created in limestone, which can be extremely porous and permeable. Limestone is eroded by chemical weathering and dissolves in water to make a solution called **carbonic acid**. Carbonic acid seeps into the ground and dissolves more limestone. The result is more fractures and joints in the limestone; these get larger with time as more water creates underground passageways and caverns.

> Speleology is the exploration of caves for fun or scientific purposes.

> The Latin word *caveo* means "beware" or "watch out."

What surface water does above the ground in the processes of erosion, transportation, and deposition, groundwater can do below. Carbonic acid trickles from a cave's ceiling, creating deposits of **calcium carbonate** in the same way that icicles are formed. The deposits that hang down from the ceiling of a cave are called **stalactites**. The drops that fall and collect on the ground to grow up are called **stalagmites**. The two may eventually meet to form a **column**.

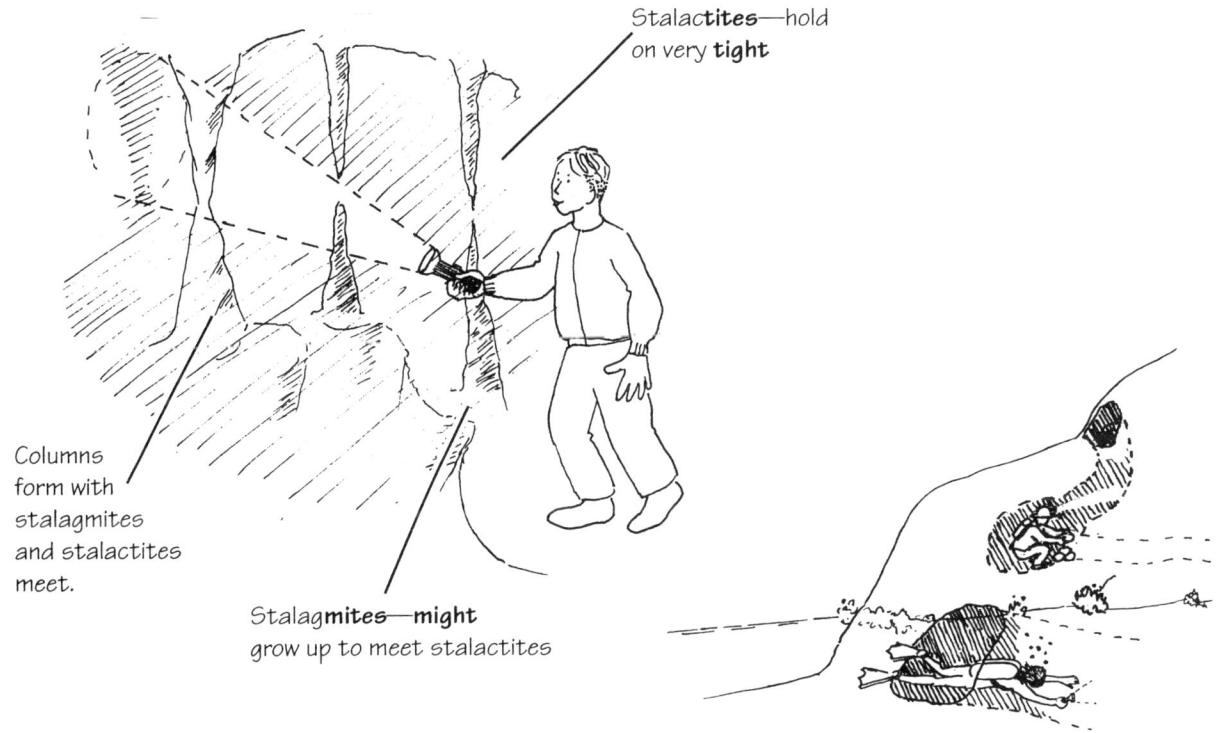

Columns form with stalagmites and stalactites meet.

Stalac**tites**—hold on very **tight**

Stalag**mites**—**might** grow up to meet stalactites

You can **spelunk** (from the Latin and Greek word meaning "cave"), or explore caves, that are above the water table or you can swim in caves below the water table in places like Florida.

MASS MOVEMENT

(to the tune of "The Boll Weevil")

The original song is a humorous take on the devastation caused by the boll weevil, a type of beetle that infests cotton fields. Carl Sandburg in *The American Songbag* writes: "A boll weevil couple, arriving in a cotton field in the springtime, will have by the end of summer, more than 12 million descendants to carry on the family traditions." [15]

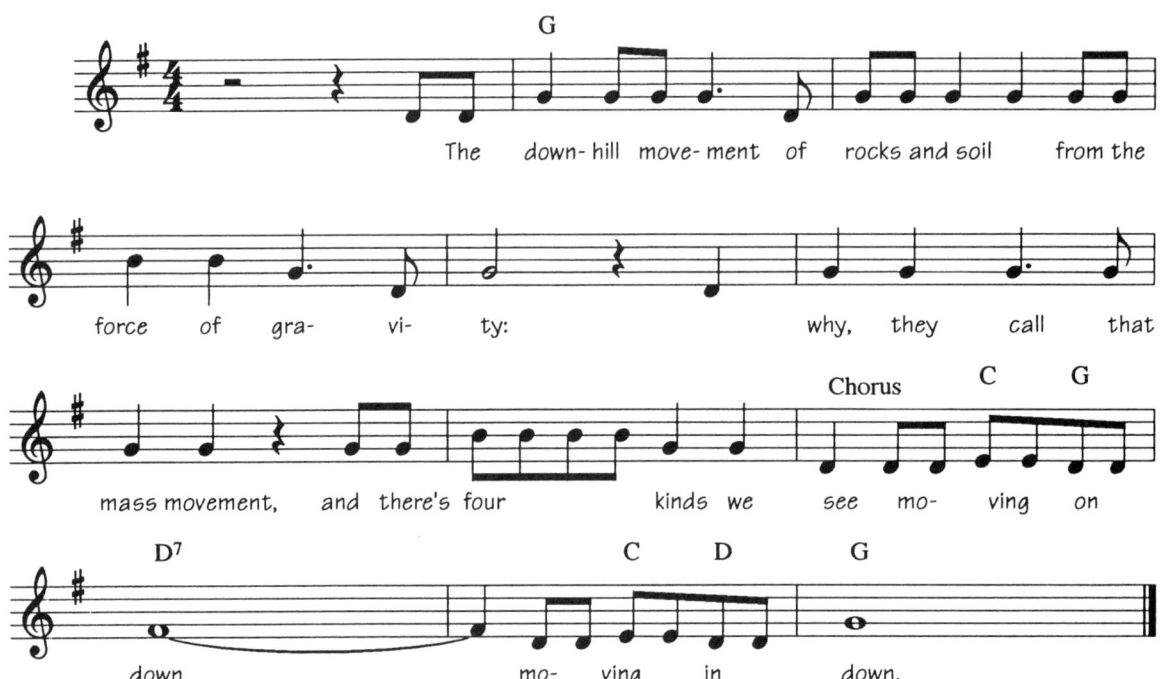

Creep is saturated ground that goes down rather slow
But if it goes down faster from more water then it's a flow
Like debris flow
Earth and mud flow

Soil and rock that goes down a steep slope when it is dry
It goes down fast, makes piles of talus; they call it a landslide
Or avalanche
Or avalanche

Mass movement with no hillside— subsidence it's name
It's what produces sinkholes of which Florida is famed
Collapsing ground
Sinkin' on down
Collapsing ground
Sinkin' on down
It's mass movement
Movin' on down
Mass movement
Movin' on down

MASS MOVEMENT

Unstable slopes of hills or mountains can be pulled down by the force of gravity. This downhill movement of a collection of rocks, gravel, sand, soil, and mud is called **mass movement**, or **mass wasting**, sometimes popularly called **landslides**. Earthquakes and flooding from rainstorms are often the cause of this process as they weaken a slope and overcome its ability to hold on to its mass. The mass movement can be at an imperceptibly slow speed over a long period of time; or it can travel quickly, creating swift destruction of anything in its path.

> Mass movements are commonly referred to as landslides, but true landslides are only one type of mass movement.

SLOPE CONDITIONS
The steepest angle which a pile of rock or soil can be stacked is called the **angle of repose**. The following conditions increase the likelihood of a mass movement to occur:
- **1-slope content**. A slope made of loose sand or other **unconsolidated** particles will be more likely to slide than a slope with compacted sediment and soil which are **consolidated**, or cemented, together.
- **2-steepness of the slope**. A steep slope will be more likely to have mass movement than a gentle slope.
- **3-water content**. The presence of water increases the chance of mass movement because it softens and lubricates the soil, and separates particles to reduce friction.

CLASSIFICATION
A mass movement is usually classified according to the amount of water in it and the speed of its descent. Four major types of mass movements are:
- **1-creep**—very slow, wet flow
- **2-flows**—faster with more water
- **3-rockfalls and rockslides**—dry rock and soil
- **4-subsidence**, or **sinkholes**—ground cave-ins.

CREEP
Creep is a mass movement of saturated, unconsolidated soil and rock particles that flow extremely slow down a slope— so slowly that changes are hard to see. A creep can take up to one year to travel less than half an inch, or one centimeter!

before

after (many years)

FLOWS

Flows are usually caused by heavy rains and move much faster than creeps. Because they flow like fluid down a slope, they cause changes that are quickly seen. More water is involved, and the amount continues to be used to classify different types of flow:

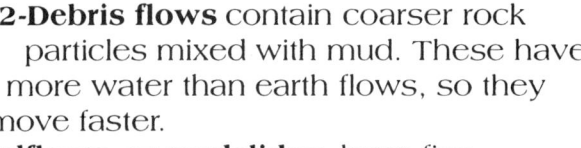
before

1- **Earth flows** consist of fine-grained particles such as sand, silt, and clay mixed with water. The slowest of the flows has the least amount of water.
2- **Debris flows** contain coarser rock particles mixed with mud. These have more water than earth flows, so they move faster.
3- **Mudflows**, or **mudslides**, have fine particles and may have some larger rocks mixed with water. They are the fastest of the flows because they hold the most water.

after

ROCKFALLS AND ROCKSLIDES

Rockfalls and rockslides, or avalanches (sometimes called landslides) refer to rapid, dry mass movements of rock and rock particles. Rock falls from a steep slope as the ground gives way when the force of gravity overcomes the slope's ability to hold onto its material. The debris deposited at the bottom of the slope is called **talus**.

1- **Rockfalls** are often caused by a joint in the **bedrock** (the solid rock underneath soil and loose rock) breaking, which releases large boulders and chunks of rock that fall down a cliff or extremely steep slope.
2- **Rockslides** or **avalanches** are loose rock and rock particles that slide down steep slopes.

Rockslide and talus

SUBSIDENCE

A process called **subsidence**, is the dropping of ground without flowing or sliding down a slope. Subsidence is a very different type of mass movement than the others, but the result is the same: large amounts of earth fall downward. **Sinkholes** may be caused by groundwater erosion weakening the surface, similar to the way that caves are created when chemical weathering dissolves limestone.

Widespread land subsidence may be caused by the overpumping of groundwater or oil which removes moisture from pore spaces (see page 67) causing them to collapse, resulting in a sinking of the ground surface. In either situation the ground becomes weak and collapses, taking with it everything on the surface: soil, rocks, horses, and barns fall into the cavity. Sinkholes are common in Florida where there are large limestone areas and many underground springs.

before

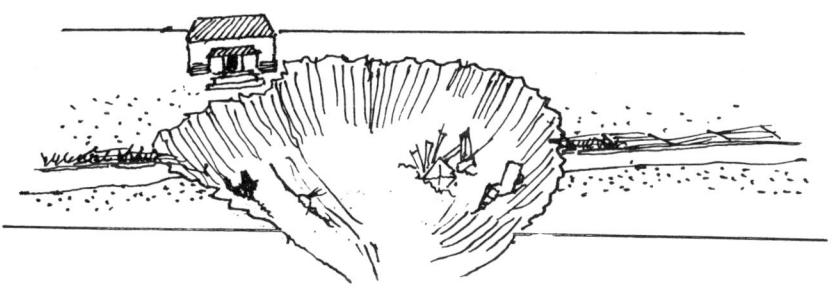

after
Sinkholes are where the ground and everything on it drops.

GLACIERS

(to the tune of "Columbia, the Gem of the Ocean")

When America's founding fathers were deciding what to name the new nation, "Columbia" was one of the possibilities. That's why the national capital is Washington D.C.; District of Columbia. Now it's remembered in this popular patriotic song.

The snow turns to firn than to ice cry- stals from ac- cumu- la- ted compacted

sn- ow, and by gravi- ty and weight (where it's thic- kest) moves the glacier in a down and outward

flow. As it moves it car- ries boul- ders rocks and gra- vel, that's drift de- po- sited as un- sorted

till; but the outwash that de- po- sits by melt- water sorts the drift by size as gravel sand and

silt. Glaciers are mas- ses of ice on the move found in mountains and high la- ti-

tudes, for- ming on- ly where snow- fall is hea- vy and the snow remains all the year

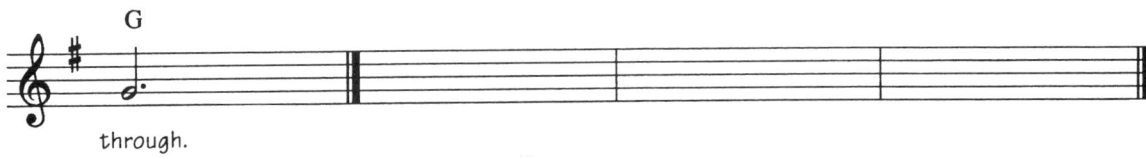
through.

Valley glaciers are formed in the mountains
    gouging out U-shaped valleys as they flow-ow
Making fjords, cirques, aretes, hanging valleys
    and striations when large rocks grind earth below
The cracks in valley glaciers are crevasses
    and deposits of sediment remain
When the glaciers start melting and retreating
    leaving fields behind of till that's called moraines

Other glaciers are continental
    which may have an ice cap, shelf, or dome
But they both shrink, it's called ablation
    and their growth is from accumulation—snow
As they move they can grind rock to rock flour
    by the pressure and weight as they erode
When this flour is mixed with the meltwater
    makes it turquoise—beautiful to behold

GLACIERS

Water, in the form of ice, covers about 10 percent of the total surface of the earth. Most is found in **glaciers**, huge sheets of ice moving over the ground. Like mass movements, glaciers are pulled downward by the force of gravity. Like rivers, they erode, transport, and deposit sediment. Glaciers dramatically change the landscape as they scrape and scratch the ground in their path; as they lift up and carry boulders, rocks, and gravel; and as they deposit them in huge fields of sediment called **moraines**.

Glaciers are formed as snow compacts and changes into ice—to do so, they need two things:
- **1-snow**—lots of it.
- **2-low temperatures** all year long to keep the snow from melting.

Instead of snowflakes, glaciers are made of firn and glacial ice, which are recrystallized snow.

Because the snow never melts but piles up, its weight exerts tremendous pressure. Fresh snow is about 90 percent air, but added weight causes it to compact and gradually change into **firn** (50 percent air) and then into ice crystals or **glacial ice** (20 percent air). This recrystallization process can take from a few years up to ten.

KINDS OF GLACIERS

Today's glaciers are remnants of those that covered as much as one-third of the earth's land surface as they once advanced toward the equator. There are clear distinctions between types of glaciers which are classified by characteristics including location and size.

- **1-Valley**, or **alpine glaciers**, are the most common, and are relatively small. Snow collects and piles up in high mountains, where glaciers form. As they grow they slowly move down into mountain valleys.
  Examples: Alaska's Mendenhall Glacier, Canada's Saskatchewan Glacier, and New Zealand's Franz Josef Glacier.
- **2-Continental glaciers** are huge sheets of ice, **ice sheets**, up to 2 miles thick in the high latitudes near the North and South poles. The ice that covers the actual poles is called an **ice cap**, or **polar cap**. (The ice cap at the North Pole is not a glacier because it's on water instead of land.) Continental glaciers

*You won't find a glacier at the North Pole—glaciers, by definition, occur on land (the North Pole is on water)—but you will find an **ice cap**.*

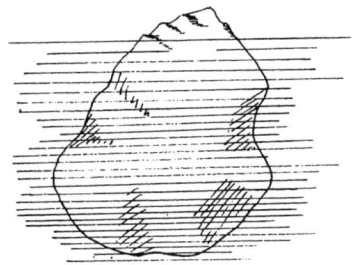

Only 10 percent of the iceberg is above the surface.

cover the entire landscape—not only mountain valleys, but the tall mountains as well. Examples: 80 percent of Greenland and 90 percent of Antarctica are covered by continental glaciers.

A continental glacier may also have an **ice shelf** attached to it, a thick layer of ice reaching into the ocean. The Ross Ice Shelf of Antarctica is a good example; it is about the size of Texas and floats on the Ross Sea.[16]

GROWTH AND SHRINKAGE

The yearly snowfall added to a glacier is called **accumulation**, and is what causes its growth. A glacier gets smaller through **ablation**, the yearly amount of shrinkage, which includes the following major processes:

1-**Melting** at the edges and sides.
2-**Calving**, or chunks of ice breaking off, forming **icebergs** where a glacier meets the sea.
3-**Sublimation**, the changing of snow directly into vapor (without first becoming water).

```
  accumulation
  − ablation
= glacial budget
```

Melting and calving are the main causes of shrinkage. If ablation is more than accumulation, the glacier shrinks; if accumulation is more than ablation, the glacier grows. The **glacial budget** is the difference between the two.

In areas where a glacier has melted, it is said to be **retreating**, though of course it can't actually move backwards (neither can a river, for obvious reasons)—the glacier is actually shrinking. A glacier is said to be **advancing** when it grows, adding more ice. Neither retreating nor advancing involves the movement of the glacier, but rather refers to the changing size of the glacier.

MOVEMENT

A glacier moves by a process called **plastic flow** in which individual crystals on the bottom melt from the pressure of the weight above—then slide past each other, causing the glacier's slow descent. Because a glacier moves by its weight and the force of gravity, it moves downward and outward from its thickest (heaviest) area. A continental glacier moves outward from the **ice dome** near its center, which may be up to 2 miles (3.2 km) thick! An alpine glacier

The speed at which a glacier moves, its **glacial speed**, increases with thicker ice and steeper slopes. The maximum speed of a glacier is about 200 feet (61 m) per year.

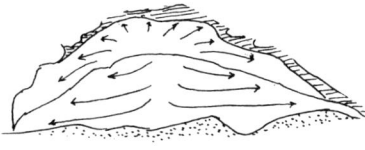

Ice dome and continental glacial movement

moves downward from a mountain or from its coldest and highest elevation. Additionally, a glacier moves fastest at the top where it is thickest, and slowest at the **terminus** (lower end) where it is thinnest. As an alpine glacier moves down the mountain, cracks called **crevasses** may form as it twists and turns with the landscape, or as it comes to an especially steep slope. Crevasses are treacherous openings for mountain climbers because they can be up to 200 feet (61 m) deep!

EROSION

As a glacier bulldozes down a mountainside and into the valley, it dislodges huge boulders and carries them along with the rocks and gravel it picks up in its path. It can grind and crush hard rock into silt and clay-size particles called **rock flour**, or **glacial flour**.

Listed below are the various erosional features and landforms carved by a glacier that are seen in areas where it has retreated.

1. **U-shaped valleys**, with flat floors and steep sides, are formed as glaciers carve the land.
2. **Fjords** are U-shaped valleys filled with water which occur along coastlines.
3. **Striations** result from boulders being dragged under the glacier along the bedrock, carving grooves that follow the direction of the flow.
4. **Cirques**, from the French word for "ring" or "circle" and derived from Latin *circus*, are deep bowl-shaped areas or depressions formed in the mountain at the head, or top, of the glacier valley.
5. **Aretes** are jagged narrow ridges between cirques.
6. **Horns** are eroded sharp mountain peaks such as the Matterhorn in the Swiss/Italian Alps.
7. **Hanging valleys** are carved by smaller **tributary glaciers**, which flow into a larger and deeper glacier. They create valleys at higher levels than the valley floor of the larger glacier.

A valley glacier only a few hundred meters wide can tear up and crush millions of tons of bedrock in a single year, enough to keep a fleet of 300 large dump trucks busy for 365 days.[17]

Glaciers carve U-shaped valleys such as the famous Yosemite Valley.

A glacier is a natural bulldozer.

Boulders, rocks, pebbles, sand, silt, and clay—anything a melting glacier once carried and then left behind is called **drift**. Glaciers may deposit this sediment in one of two ways:
    1-by melting ice
    2-by flowing water, called **meltwater**, which is a stream within and underneath a glacier.

LANDFORMS FROM TILL
—DEPOSITS MADE BY ICE

Sediment which ice deposits is called **till** and is not stratified but unstratified and unsorted, with the largest rocks and finest silt all mixed up together. Large boulders left behind as a glacier melts are called **erratics** because they are not like the rocks found naturally in the area.

Moraines are large fields or areas of the accumulated till. The names and kinds of moraines are determined by their location:
    1-end or **terminal moraine** is at the front or lower end of the glacier.
    2-lateral moraine is at the sides.
    3-medial moraine is where two glaciers meet.
    4-ground moraine is beneath the glacier.

Drumlins are gently sloping hills of till in the direction of the ice flow.

LANDFORMS FROM OUTWASH
—DEPOSITS MADE BY MELTWATER

Sediment laid down by streams of meltwater is sorted by size and stratified like the deposits of any stream. If drift has been carried and sorted by meltwater, the deposit is called **outwash** because it has been "washed out" of the glacier.

Kames are low mounds of sand and gravel that form a fan or delta-like deposit near the lower end of a glacier.

Eskers, from the Irish word for "ridges," are found only in ground moraines because they are formed by streams of meltwater as they flow through the bottom of a glacier. The deposits of sand and gravel create ridges at the stream's banks, and are exposed when the glacier recedes.

Kettles are formed from large chunks of ice left behind and buried in outwash that create steep, bowl-shaped holes as they melt. Thoreau's Walden Pond in northeast Massachusetts, is an example of a **kettle lake**, a kettle that has filled with water.

A WORD ABOUT PERMAFROST
Though **permafrost** may be as cold as a glacier, it's not a glacier but ground that remains frozen all year around.

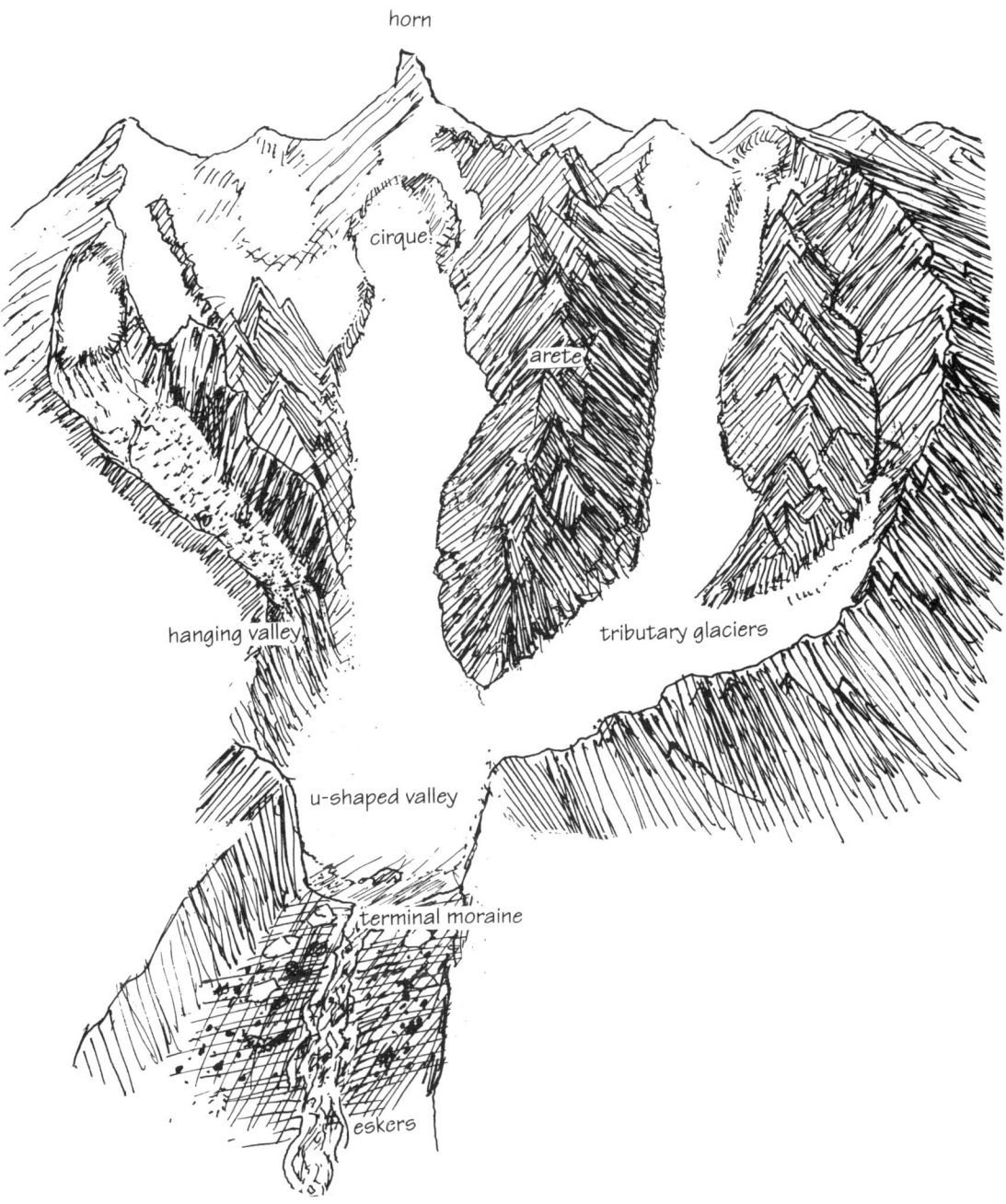

WIND EROSION

(to the tune of "Blow Ye Winds in the Morning")
This joyful melody is from a song of the early whalers of the Northern seaboard.
'Tis advertised in Boston, New York and Buffalo:
500 brave Americans are sailing for to go,
Singing blow ye winds in the morning.

In regions that are dry and where the sur- face of the ground is lack- ing ve- geta- tion but loose

par- ti- cles of rock are found; and loose ma- ter- i- al blowing can be much like sand- bla- sting; and

wind e- ro- sion can make rock for- ma- tions that are in- ter- es- ting. Blow- ing wind wears rocks away

blo- wing wind hi ho. A- bra- sion and scour- ing from the wind e- rodes.

In the desert where all loose material is blown away
A desert pavement's all that's left from a process called deflation
Boulders may get polished faces, but they're seldom round
The windward side is flattened, meaning sharper angles will be found
Chorus

Wind erosion varies with how long and hard it blows
And the size of grain, whether sand or silt or loess, but
When wind is slowed down because of plants or topography
Then sand dunes can be found there, but the silt and dust in clouds keep blowing
Chorus

WIND EROSION

Sandblasting is the process of blowing sand out of a nozzle under great pressure. It is used to remove old paint from sides of buildings, cars, and bikes. It's similar to the process at work in wind erosion, which scours and also removes surface particles. Wind then lifts these weathered rock particles and carries them along to be deposited elsewhere. One spectacular result of wind erosion is the exposure of volcanic rock formations (discussed on page 33) where wind has eroded softer rock, leaving harder rock behind. Wind carves its own unique geologic features, such as those found in Zion National Park. Natural bridges, found in arid (dry) regions, are often carved by wind erosion.

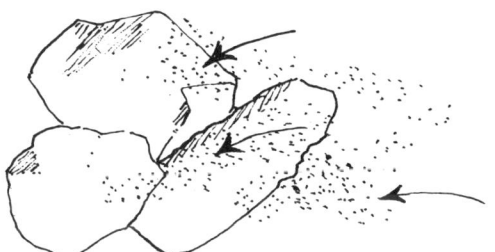

Arches carved by the wind

Wind moves horizontally along the surface of the ground. The way it erodes is similar to water in that the **turbulence** of the current picks up and carries fine rock particles and deposits them in ripples and dunes. Wind erodes quite differently from water in that wind (unless in a tornado or hurricane) cannot carry or move large objects.

Even though the wind can't pick up large rocks or boulders, it reshapes them as it scours them with blowing particles, such as sand.

FACTORS AFFECTING RATE OF EROSION
Wind can carry thousands of tons of soil, as in the dust storms of the 1930s that carried away the topsoil of the Great Plains, creating the "Dust Bowl." Wind can carry volcanic ash and dust great distances. When Mt. St. Helens erupted in 1980, the cloud of dust and ash shot up several miles and was blown hundreds of miles east. The amount of dust wind can carry varies greatly and is determined by several factors including:

1-size of rock particles—wind can carry the smallest particles such as clay (dust) and silt more easily than it can carry larger particles such as sand. **Loess**, a combination of these three, is deposited by the wind and creates fertile soil, good for growing crops (see page 55).

2-**the strength of the wind**—strong winds carry more than quiet, gentle breezes; they also can carry larger particles, such as sand.

3-**duration**—winds that blow for long periods are able to carry more particles than short gusts.

CAUSES OF HIGH WIND EROSION

What makes some areas and some soils more prone to wind erosion than others? Why was the topsoil lost during the great dust storms of the Dust Bowl era? Several causes of high wind erosion are:

1-**lack of moisture**—dry ground is necessary for wind to be able to lift up sand and dust because moisture makes particles stick together. Drought was the main cause of the 1930s dust storms.

2-**lack of vegetation**—roots help hold soil together, and vegetation slows or breaks the speed of the wind. Both unplanted farm fields and desert areas with little vegetation make the ground vulnerable to wind erosion.

3-**agriculture methods**—overgrazing by cattle weakens or destroys vegetation. Plowing furrows in line with the direction of the wind causes more soil to blow away.

4-**thin soil**—soil that lacks depth doesn't retain moisture and so erodes quickly.

5-**strong winds**—as mentioned above, strong winds cause more erosion. The Great Plains of the Midwest are known as having strong, almost constant, winds.

The result of all of the above—environmental and economic ruin.

The Dust Bowl

LANDFORMS CREATED BY WIND EROSION

Sand dunes form as the wind slows and sand grains moved by the wind are deposited. Sand moves by **saltation**, a process in which it bounces forward in short jumps as the wind blows it along. The side of the dune facing into the wind (windward side) has a gentle slope, and the side facing away (lee side) has a steeper slope. Sand dunes move, or migrate, as wind pushes sand over the crest and deposits it on the other side.

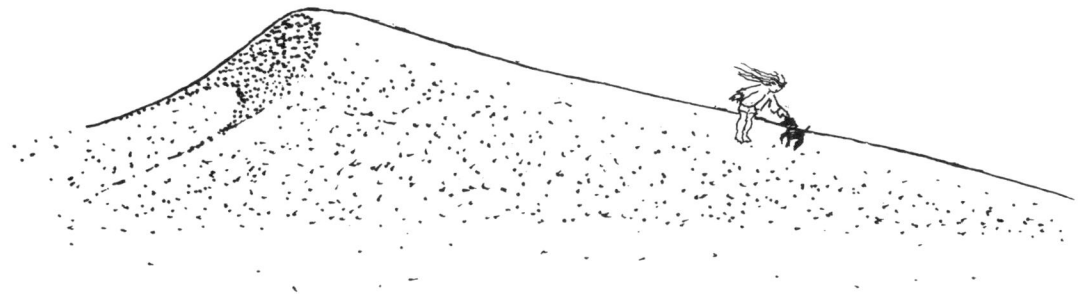

The wind moves sand up and over the crest of the dune.

Deflation is the lowering of the ground level from the winds' scouring action, which erodes the smaller particles, such as sand. A **desert pavement** consists of compacted smooth pebbles and larger rocks left behind from the deflation process. In the Middle East, huge areas of desert pavement play a major role in controlling wind erosion. The hard pavement (on which cars and trucks can even be driven) keeps the sand below it from blowing away and creating sand dunes that migrate with the wind.

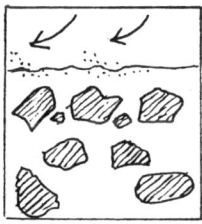 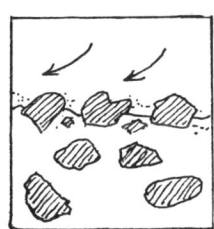  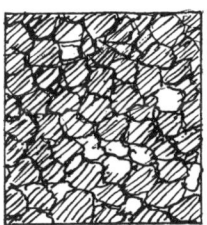

Desert pavements become compacted as small particles such as sand are blown away.

SOIL

(to the tune of "My Grandfather's Clock")
This melody is from a loving, sentimental song from the 1800s about a grandfather, his life, his clock—and the passing of time.

For soil to be formed other factors to recall,
   include what kind of rock it came from
Called parent material with the kinds of minerals
   that affect what the soil will become
Climate (temperature and rain)
   and the slope of the terrain
Solid particles of sand, silt, and clay
With pores that are either filled with air
Or with water at any one time

A profile and colored layers that it shows:
   are horizons of A, B, and C
Humus from plants decomposed makes the topsoil
   of A more fertile than B
The minerals on top leaching down will likely stop
   at the C-horizon (partially weathered rock)
Laterites—climate wet
Pedalfers—temperate
Pedocals where the climate is dry

SOIL

Soil is the result of physical and chemical weathering, and is composed of **inorganic** (non-living) minerals and rock particles and **organic** (living or once-living) plant material. Soil can measure anywhere from a few inches to several feet deep. Many of us might not think about the importance of soil, but farmers and gardeners know the value of fertile soil for growing healthy plants. By looking back at the history of humans, we find that soil has been of great importance.

Agriculture, the growing of plants for people to use, or to feed their animals, was the foundation of civilization. The plants we grow, called crops, are important to provide food, clothing, oils, and shelter for people—and these plants grow in soil.

Agriculture
Grains or cereals
   wheat
   rice
   corn
   oats
Fruits
   apples
   bananas
   citrus
   nuts
Vegetables
   potatoes
   carrots
   lettuce
Fiber
   cotton
   flax (linen)
Oil
   sunflower
   olive
Feed (for animals)
   pasture grass
   alfalfa
   grains
Trees
   lumber
   fiber for paper products
Ornamentals
   flowers
   shrubs

*Soil science is also called **pedology**, which comes from the Greek word pedon, meaning "ground." Soil science is a specialized study that involves knowledge of other sciences, including geology, biology, and chemistry.*

WEATHERING OF ROCK INTO SOIL

Weathered rock particles make up, or compose, at least 80 percent of soil. Soil takes a long time to form, and the amount made and its characteristics are affected by several factors.

1. **Parent material** is the original rock the soil is weathered from, and it affects the characteristics of the soil it makes. Different parent materials will form very different soils. For example, soil made from limestone will be different from that made from shale.
2. **Climate** affects the speed of soil formation the same way it affects the weathering of rocks. Warm climates with lots of rain create soil faster than colder, drier climates. Climates with great changes of temperature also weather rock quickly and so create soil faster than more temperate climates.
3. **Topography** includes the slope of the terrain and how much direct sunlight it receives, which in turn affects the temperature and speed of weathering. There isn't much soil on steep slopes because gravity causes mass movements. Instead, as mountains erode, their particles are deposited at their base, on more level ground.
4. **Organisms** such as bacteria and fungi cause faster soil formation because they help break down plant and animal materials which then get added to the weathered rock particles of sand, silt, and clay.

Dirt is a commonly misused word. Dirt is the substance that gets on clothes, skin, and floors. We usually end up washing it off.

ORGANIC MATTER AND ORGANISMS

Rotting and decaying plants release minerals, (remember, there are several thousand minerals and only a very few form rock) to the soil. What does not break down right away becomes **humus** (HYOU mus). Humus and animal material are types of organic matter—a small but important part of the soil. The word "organic" means "containing carbon" or "derived from living organisms." These organisms and substances found in soil can be divided into:

1-those which supply organic matter— mostly plants (living and dead) and some animal material.

2-those which break down organic matter— or feed either on that material or directly on other living things. Bacteria and fungi are the most important members of this group.

*As decomposers break down plant material—what happens in a compost pile—carbon and other minerals are released back into the soil, making it rich, dark, and **fertile**, good for growing plants.*

Both living and dead plants are a source of food for many different organisms, and those organisms are themselves a source of food to others in a complex soil ecosystem. The plants we see growing above ground are the **producers** in that ecosystem, and the animals that eat them are **consumers**. But the most numerous of the soil organisms are the **decomposers**, the fungi and bacteria, which get their energy from breaking down the other organisms.

What lives in the soil:
Producers:
  Algae
  Plants
Consumers:
  Rodents and insectivores (mice, gophers, ground squirrels, moles, shrews, prairie dogs)
  Arthropods (insects such as ants, beetles, centipedes, sowbugs, millipedes, mites)
  Mollusks (slugs and snails)
  Earthworms
  Nematodes (microscopic round worms)
  Protists (single-celled animals)
Decomposers:
  Fungi (molds, mushrooms, rots, etc.)
  Bacteria

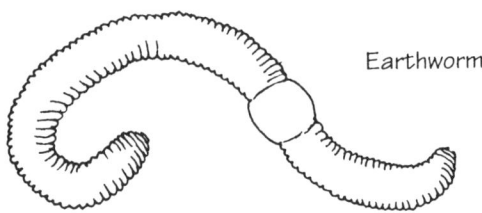
Earthworm

A benefit that organic matter and organisms add is increased pores, or spaces, between the solid particles of soil. Soil that's too compacted has smaller and fewer pores and will not be as healthy for plants. Looser soil, developed mostly by organisms' movements, allows water, air, and nutrients to easily reach the roots. Looser soil also helps roots easily penetrate deep into the soil as they grow.

At any particular time, these pores are filled with either air or water—how much of each and for how long determines how well plants grow in that soil. When you water your plants, water fills the pores—the plants grow as they absorb the water from around the roots. If you forget to water, air fills the pores, the soil dries, and your plants wither and dry up.

Soil

SOIL PROFILES

Soil is formed in horizontal layers called **horizons**. Horizons often have different colors and textures, and vary both in composition (what they are made of) and formation (how they are made). Horizons are labeled with capital letters to identify them and their distinctive characteristics.

> **A horizon** is the top layer and is composed of humus, sand, silt, and clay, roots, and many soil organisms. It is where most plants grow.
>
> **B horizon** is just below the A horizon and may be made of sand, silt and clay, a little humus, and maybe a few roots.
>
> **C horizon** is partially weathered rock.
>
> **D horizon** is bedrock.

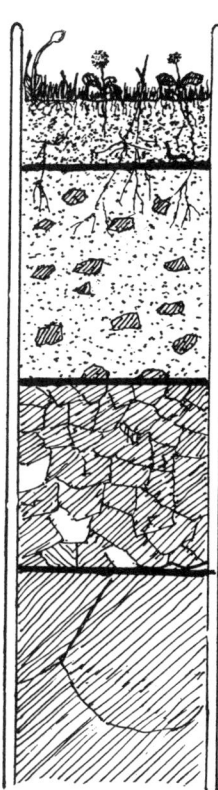

A soil profile

Minerals from the dissolved load of rivers, and minerals released from the decomposition of plants near the surface, are what make soil **fertile**, or good for growing plants. But these minerals needed for plant growth, or **nutrients**, do not always remain in the A horizon. In areas with ample rainfall, water seeps down and carries with it minerals from the A horizon to the B horizon in a process called **leaching**. This causes the A horizon to be less fertile.

A soil **profile** is a vertical slice of soil and includes all the horizons. It represents a side view of soil—what you see when a hole is dug or a road is cut into a hill.

> Soil classes are based on texture—the proportions of sand, silt, and clay. **Loam** is the ideal balance of sand, silt, and clay for agriculture. The name of the soil class reflects the proportions of the particle size.
> - loamy sand
> - sandy loam
> - loam
> - silt loam
> - silty clay loam
> - clay loam
> - clay

SOIL GROUPS

Differences in climate cause different amounts and types of weathering and so create soils with various compositions and characteristics. These differences can be classified into the following soil groups.

> **1-Laterites** are created from warm, wet climates such as rainforests. Weathering is rapid, creating deep, thick soil. But because there is so much precipitation, nutrients leach from the A horizon, leaving soil good for native plants, but not good for growing grain, fruit, or vegetable crops.
>
> **2-Pedalfers** are formed in temperate climates (moderate temperatures and rainfall) such as the eastern United States and Canada. The soil is thick and rich—the most fertile kind for growing crops.
>
> **3-Pedocals** are formed in dry climates, such as deserts. Weathering is slow, causing soil to be thin and unproductive.

Plowing helps return humus and minerals to the top of the soil.

TOPOGRAPHIC MAPS

(to the tune of "Life on the Ocean Waves")
"Life on the Ocean Waves" is probably one of the most well known sea shanties.

Brown lines will point upstream when crossing rivers or creeks or valleys
And as confusing as it may seem, the branching of blue tributaries
Will point where water flows
And one more thing to know is...
    The U.S. Geological Survey is a governmental agency
    That surveys, maps, and studies landforms and geologic activities
    And puts benchmarks at certain locations, and
The U.S.G.S. has colors on their topo maps are seen
Black is for what's man-made; water is blue; and plants are green

TOPOGRAPHIC MAPS

Different types of maps—
  topo maps
  road maps
  political maps
  physical maps
  relief maps

Maps help us understand where things are in relation to other places from a bird's-eye view, the perspective of looking down from above. We plan and find our way as we drive or bike by using road maps; we learn of the sizes and locations of cities, states, and countries from political maps; and we see geographic features in color on physical maps, or standing out in three dimensions on relief maps. We study the shape of the land with **topographic maps**, often called **topo** (TOE poe) **maps** for short. These maps show the shape of the land in detail by representing three dimensions on a two-dimensional, or flat, page.

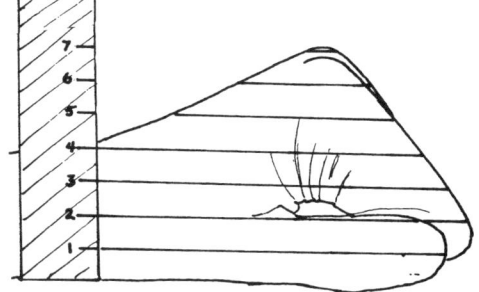

Straight lines made on a fist are seen to be the same distance apart when viewed from the side.

To be able to read topo maps, you need to understand that the lines you see are symbolic, not anything you will see from an airplane or marked on the ground. What you will find on topo maps are:

1-**legend**—brief description to understand the map
2-**scale**—the ratio, or proportion of distances on the map to distances on land
3-**contour lines**—the lines representing the shape of the land
4-**contour interval**—the distance between contour lines
5-**index contour lines**—labeled with the elevation

Topographic maps represent the surface features of an area.

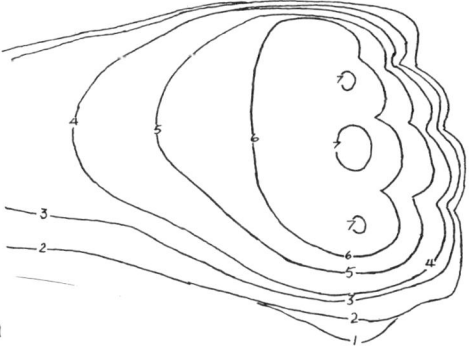

These same lines, when viewed from above, are seen to be closer together when the slope is steep and further apart when the slope is gentle. This demonstrates how contour lines show three-dimensional elevations on a flat two-dimensional map

Altitude is the height in the air (as the altitude of a flying plane). **Elevation** is the height of the ground (as the elevation of a mountain top).

The **legend**, or **map key**, includes explanations and information such as a **scale** that compares the distances on the map to those on the land. For example: one inch on the map may represents one mile on land, or one centimeter may equal one kilometer.

TOPO MAP LINES

The **contour lines** show the shape of the land because they represent connecting points having the same **elevation**, the height above sea level. They make an outline of one single elevation on a landform. For example: contour lines become small circles on hilltops or mountains, and circles farther apart at lower elevations. Generally, the contour lines do not branch or fork or even touch. When lines are close together on a topo map, that means the landform's slope is steep; when the space between lines is greater, the slope is more gentle. An almost vertical slope would have lines almost on top of each other; a very gradual slope would have widely spaced lines.

> When contour lines are:
> 1-close together, they show that a slope is steep.
> 2-farther apart, they show a gentle slope.
> 3-even farther apart, they show a valley or flat area.

Every fifth line is darker and thicker; these are called **index contour lines** which are labeled with their elevation, which allows you to figure out the elevations of the four lines in between by dividing the distance by five.

> Index contour lines are marked with the elevation so you can calculate the elevation of the contour lines inbetween them.

The distance between contour lines is called the **contour interval**, and it represents the distance up or down in elevation in regular increments. That information is usually given in the legend. If the contour interval is 20 feet, you could estimate the elevation of a particular point by finding the elevation of the lines on each side of it. For example, point "X" could be estimated to be at about 30 feet because it is halfway between the 20-ft. and the 40-ft. lines.

> The contour interval is the distance between contour lines and is usually given in the legend in either feet or meters.

As shown on the topo map of the fist, the contour lines between fingers, which could represent river canyons, look like they point upstream. Water actually flows the opposite way contour lines point. But when you look at a river system with **tributaries** (creeks and streams flowing into a larger river, see page 64), you can also interpret which way the water flows by finding a place where a tributary joins a larger stream or river. That point is called a **confluence**, and the smaller angle, like an arrow, will point downstream.

The topo maps produced by the **U.S. Geological Survey** (or **USGS**) use these colors, lines, and symbols:
      1-black—man-made structures (houses, buildings, smaller roads, etc.)
      2-brown—contour lines
      3-blue—water
      4-green—vegetation
      5-red—boundaries and main roads

Topographic Maps

U.S. GEOLOGICAL SURVEY

The United States Geological Survey was established when the country was young and unexplored. The Department of the Interior needed accurate surveys as settlers moved west and claimed land, and as boundaries between counties and states were established. Using surveying equipment and techniques, and mathematics, the USGS made maps of the growing country, and placed markers, called **benchmarks**, on tops of hills and mountains, and at the locations corresponding to the corners of the grid squares on their maps.

In modern times the USGS continues to play a valuable role with its geological research of volcanoes, earthquakes—and more. We strongly encourage you visit The Learning Web part of the USGS website to find out more about plate tectonics, landforms, earthquakes, volcanoes, erosion, rocks and minerals, water, glaciers, soil—and topographic maps!

The topo map below shows the elevation and features of this landscape.

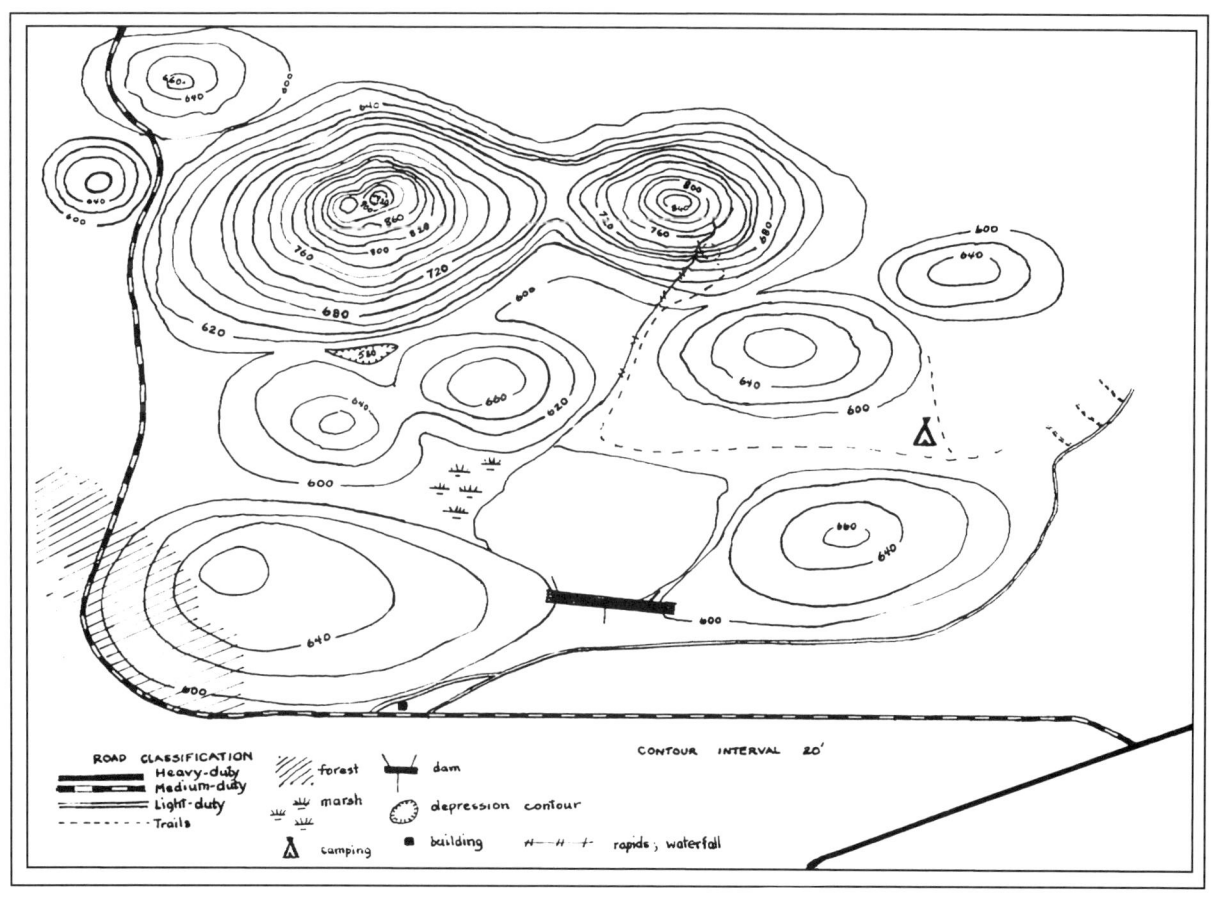

CONCLUSION

So now you know much more about geology
So much more than when you first began
With the terms and the concepts and the geologic processes
The earth around you'll better understand!

APPENDIX

Song lyrics
Notes
Bibliography
Index

AN INTRODUCTION TO GEOLOGY

(to the tune of "Battle Cry of Freedom")

With our need to understand, to find meaning and make sense
Searching the past for explanations
We interpret, when we can, what we find as evidence
Searching the past for explanations

> Chorus:
> We take the stories of how, why, and when
> Into our theories, til questions come again
> Information we acquire gives the knowledge we desire
> Searching the past for explanations

So with history it is, both of humans and the earth
Searching the past for explanations
In geology we know all the effort makes it worth
Our searching the past for explanations

> Chorus:
> Landforms and fossils, from mountains to old bones
> Are geologic puzzles and riddles in the stone
> So we deduce and we infer what we think had to occur
> From searching the past for explanations

The geologic processes that we see today
Searching the rock for explanations
Some think to have occurred so long ago in the same way
Searching the rock for explanations

> Chorus:
> "Uniformitarianism" things change the same
> But "catastrophism" says major events came
> An upheaval: an earthquake, or eruption, or a flood,
> We're still searching the rock for explanations

Deposits of new sediment are on top of older layers
Searching the rock for explanations
So "relative age" means just the age, but not in years
Searching the rock for explanations

> Chorus:
> "Superposition" means older rock below
> Comparison by age but it doesn't help us know
> The absolute age; so they estimate the date
> From searching the rock for explanations

PLATE TECTONICS
(to the tune of "Wabash Cannonball")

Continental plates of crust of thinner lithosphere
And heavy oceanic plates above asthenosphere
Tectonic plates seem all to move and the continents to drift
And in valleys and the ocean floor the spreading causes rift.

The Mid-Atlantic Ridge is from divergent boundary plates
New crust forms and pulls apart where they separate
But volcanic vents called hot spots in New Zealand and Galapagos
Hawaii and in Iceland, and in Yellowstone

Tectonic plates converging, oceanic plate subducts
Under lighter continent—bends down then it melts
Pacific Rim's a "ring of fire": volcanoes and earthquakes
Convergent boundaries where you'll find those subducting plates

Tectonic plates instead may move more from side to side
At a transform boundary, strike-slip fault; at lateral faults they slide
California's San Andreas fault is where you'll find
One of the examples of movement of this kind

FOLDING, FAULTING, AND INTRUSION
(to the tune of "Washington Post March")

Although many mountains are volcanoes
    There are forces that will change
    The crust of Earth into a range
Of mountains which are not volcanic
Tectonic forces cause the folding and the faulting of the crust

And when these forces do cause folding
    Creating A-shaped anticlines
    And U shape of synclines
Twisting, folding, squeezing, pulling
Movement that's along a fracture is a fault and not a joint

And with faults there are three main kinds
    Direction of force
    Will vary, of course
The footwall and the hanging wall
If moving apart; the tension causes a normal fault

But if the force is the other way
    Compression will push
    A fault in reverse
But then a shear or a sideways force
Can make a strike-slip fault like San Andreas

Intrusive igneous rock bodies
Where magma underground slowly cooled into rock
    Can be as large as an entire mountain system
    In batholiths, or smaller in a stock

A dike is magma that hardens after cutting 'cross the layers
A sill is where magma squeezed between
    And a laccolith is like a sill except that it's thicker
    Pegmatite is where large minerals are seen

So now you know some structural mountain forms
From folding, faulting, intrusion, is uplift
    But if the landforms are formed instead from erosion
    Dissected mountains are what is left

EARTHQUAKES

(to the tune of "Little 'Baccy Box")

We've talked already about tectonic plates
But now it's time to discuss earthquakes
Earthquakes are the release of energy
And the science of earthquakes is seismology
Seismo means "shaking," like in seismology

Plates converging, all the forces and the stress
From the pushing and the sliding, all the tension you may guess
Land begins to shake 'til releasing makes it stop
Focus in the mantle, epicenter's on the top
Focus in the mantle, epicenter's on the top

Seismographs at the seismic stations
Seismograms record the wave vibrations
Measure magnitude or intensity
Relative amount of released energy
Richter Scales and amount of energy

Deep primary waves are longitudinal
They stretch and compress and they push and they pull
Secondary—s waves, slower don't you know
Side-to-side or up or down they go
Side-to-side or up or down they go

And surface waves all along the surface roll
Together with the s waves they really take a toll
Shaking buildings down and shaking everything
Blame these kinds of waves for all trouble that they bring
Blame these kinds of waves for all trouble that they bring

VOLCANOES

(to the tune of "The Minstrel Boy")

There's Mt. St. Helens, Hood, and Fuji
In the ring of fire—they are volcanoes.
Vesuvius the mountain in Pompeii,
Krakatoa—they all are volcanoes.

    Chorus:
    The crust that moves as tectonic plates
    Colliding at subduction zones relates
    To magma rising to the surface crust
    Making lava through vents of the volcanoes

Mafic magma features they include
Flowing lava, high temperature is fluid
Basalt of lava fountains, falls, and tubes
Cinder cones, or most common—shield volcanoes

    Chorus:
    Of flowing lava the main kinds are two:
    Pahoehoe lava is the one that's smooth
    And aa is so rough and broken
    Coming from Mauna Loa, the volcano

Felsic magma temperature is low
Make the pyroclastic flow from the volcanoes
And with intermediate of flow and ash
Make composite, or the stratovolcanoes

    Chorus:
    Fissure eruptions may form plateaus
    Gasses and steam come from the fumaroles
    Lava, ashes, when they're mixed and blown:
    Spew from craters and vents of the volcanoes

MINERALS
(to the tune of "Invalid Corps")

Minerals are inorganic solids found in nature
With certain properties like crystal form, cleavage, and fracture
And hardness, density or heft, color, streak, and luster
Constant composition of their chemical's a feature

>Chorus:
>Minerals are inorganic solids found in nature
>Constant composition of their chemical's a feature

Minerals on earth are found as element or compound
And out of several thousand only twelve are commonly found
Rock made out of minerals with silicon are silicates
Other groups are oxides, carbonates, sulfates, and phosphates
Chorus

Minerals called precious metals: silver, gold, and platinum
Industrial-type metals: iron, copper, tin, titanium
Nickel, manganese, cobalt, lead, borax, molybdenum
You'd be quite surprised at all the products that contain them
Chorus

Gemstones are the minerals rare and of high value
Brilliant, durable, colorful jewelry they're made into
Diamonds, emeralds, sapphires, and rubies are examples
Turquoise, malachite, and jade—a few of opaque samples
Chorus

IGNEOUS ROCK

(to the tune of "Anchors Away")

Magma cools into rock, igneous rock
It's what's in both the other—sedimentary, metamorphic
Classification is by chemicals.
And texture from the crystals, large, or small or none at all

Igneous rock can be one of two kinds
Intrusive starts as magma and cools slowly underneath the
ground—like in granite, large crystals explain
Why this abundant rock has coarser texture, it's the coarser grain

Igneous can also be extrusive rock
When magma becomes lava at the surface from volcanoes
Where it is quickly cooled, crystals are small
Finer grain the finer texture—the most common is basalt

Magma can turn to glass when it cools fast
No crystals in the dark obsidian, pumice bubbles within
So for texture there are these three kinds
With crystals large, or crystals small, or with crystals not at all.

SEDIMENTARY ROCK

(to the tune of "Bonny Blue Flag")

Chorus:
Rocks, rocks, sedimentary rocks:
Clastic, chemical, organic sedimentary rocks
Rocks, rocks, sedimentary rocks:
Clastic, chemical, organic sedimentary rocks

Rocks are broken down and as sediment carried away
As boulders, cobbles, pebbles, gravel, sand, or silt or clay
When they are deposited in layers they may all
Be cemented in conglomerate, sandstone, or shale
Chorus

Minerals that were dissolved in water at one time
Precipitate or from evaporation are left behind
Rocks are formed from some of these chemical sediments:
Gypsum, rock salt, compact limestone, chalk, and chert or flint
Chorus

Organic limestone comes from hard remains of animals
Peat from plants decayed compresses to lignite then to coal
All three kinds of sedimentary rocks are valuable
In building and construction and in things industrial
Chorus

METAMORPHIC ROCK
(to the tune of "Just Before the Battle, Mother")

The kind of rock called metamorphic
Is made of rocks that have been changed
Perhaps in recrystallization
A change in how they are arranged
Or in size or shape of crystals
Affecting texture and the grain
Or change in mineral composition
Which means the rock that was is not the same

If you ask what cause these changes
Making rock so hard and dense
I'll tell you it's from heat and pressure
From the moving continents
Heat from intruding nearby magma
In mountain building, or the weight
Of tons of sedimentary strata
Or faulting, folding from tectonic plates

Foliation looks like layers
Shale or siltstone becomes slate,
With more heat and with more pressure
Hardens to phyllite, or schist, or gneiss,
The metamorphic rock that shows no layers
Nonfoliated rock has layers not seen:
Limestone hardens into marble,
And basalt to serpentine

WEATHERING OF ROCKS
(to the tune of "Wearing of the Green")

In the changing of the landscape, a most important thing
Is how the rocks are broken down; we call it weathering
There are two kinds of weathering: one is physical
Decomposition is the other, known as chemical
And four things that affect the rate of all weathering
Composition of the rock; the climate (temperature and rain);
And topography's exposure to the wind and rain and sun
But vegetation may provide the rocks some protection

There are four kinds of weathering we say are physical:
Disintegration happens, and it is mechanical:
Exfoliation is the flaking off of rocks in sheets
When rock expands resulting in surface cracks and joints
Another is frost action: freezing water expands cracks
And from growing plants the prying roots do slow, destructive acts
And finally there's abrasion from physical contact
When edges sharp are rounded, by both grinding and impact

Now to all those kinds of weathering with chemical compare
Which decomposes rock by oxidation from the air;
Or hydrolysis from water; or by acid from the rain;
Or from the lichen growing; or from plants decayin'

HYDROLOGY AND EROSION
(to the tune of "Yankee Doodle Dandy")

Water moving and effects it has: the science of hydrology
Includes the water cycle as it moves through air to the ground to the sea
From oceans much evaporation; clouds are condensation in the air
Rain and snow's precipitation; ground water from infiltration,
Surface runoff flows towards the sea.

Volume is amount of water, velocity is speed as water flows
Load's amount of sediment carried; depending on volume and speed
Capacity is the ability of water to carry sediment
Erosion is the movement of the sediment by runoff
Deposition drops the load when water slows

The amount of runoff is determined partly by topography:
Steeper gradient or slope will mean, the faster the runoff will be
Rate of precipitation affects runoff, vegetation slows the water down.
Thicker humus and the looser soil will also
Make more water soak into the ground

WATERWAYS AND EROSION

(to the tune of "Shenandoah")

When water flows, it always follows
Downhill paths of least resistance
The runoff finds the shortest distance
Obstructions found
It will flow around
But cuts through soft ground

Erosion is the process that moves
Rock and soil—the land reshaping
When water, wind, and glaciers carry
And wear away—
Erode away
Highlands into lowlands

An area drained by a river
And its branching tributaries—
A watershed is separated
By a divide
And on each side
Water flows in two directions

A river's source is the headwaters
Making channels and V-shaped valleys
Straight, steep, and wild; rapid erosion
Rivers flow
While they erode
The ground beneath them

Volumes rise, high deposition
Meanders on the flat wide floodplains
Continues to mouth or base level
Forms deltas there
And that is where
It flows into the ocean

GROUNDWATER
(to the tune of Handel's "Water Music Suite")

Water underground
Flowing underground
Through rock that's porous and permeable
Porous rock is like a sponge, and if the pores are connected rock is permeable

Water passes through
Slowly continues
Seeping down through the aeration zone
To the water table, where the pores are full of water—saturation zone

A permeable layer
Is an aquifer
And a spring is groundwater that comes
To the surface naturally
And water tables may be
At the surface in a marsh, lake or stream
Artesian wells from pressure rise above aquifers

Water underground
Flowing underground
Through caves and caverns it erodes
Minerals from chemical weathering are dissolved and may be deposited—

On hanging stalactites
Onto stalagmites
And in a column they may join
These calcium carbonate
Formations may be great
All this from groundwater

Water underground
Slowly flowing
Underground

MASS MOVEMENT
(to the tune of "The Boll Weevil")

The downhill movement of rocks and soil from the force of gravity
Why, they call that mass movement, and there's four kinds we see
Moving on down. moving on down

Creep is saturated ground that goes down rather slow
But if it goes down faster from more water then it's a flow
Like debris flow, earth and mud flow

Soil and rock that goes down a steep slope when it is dry
It goes down fast, makes piles of talus; they call it a landslide
Or avalanche, or avalanche

Mass movement with no hillside— subsidence its name
It's what produces sinkholes of which Florida is famed
Collapsing ground, sinkin' on down
Collapsing ground, sinkin' on down
It's mass movement, movin' on down
Mass movement, movin' on down

GLACIERS
(to the tune of "Columbia, the Gem of the Ocean")

    Chorus:
    Glaciers are masses of ice on the move
    Found in mountains and high latitudes
    Forming only where snowfall is heavy
    And the snow remains all the year through

The snow turns to firn then to ice crystals
    from accumulated, compacted snow-ow
And by gravity and weight (where it's thickest)
    moves the glacier in a down and outward flow
As it moves it carries boulders, rocks, and gravel
    that's drift deposited as unsorted till
But the outwash that deposits by meltwater
    sorts the drift by size as gravel, sand, and silt
Chorus

Valley glaciers are formed in the mountains
    gouging out U-shaped valleys as they flow-ow
Making fjords, cirques, aretes, hanging valleys
    and striations when large rocks grind earth below
The cracks in valley glaciers are crevasses
    and deposits of sediment remain
When the glaciers start melting and retreating
    leaving fields behind of till that's called moraines
Chorus

Other glaciers are continental
    which may have an ice cap, shelf, or dome
But they both shrink, it's called ablation
    and their growth is from accumulation—snow
As they move they can grind rock to rock flour
    by the pressure and weight as they erode
When this flour is mixed with the meltwater
    makes it turquoise—beautiful to behold
Chorus

WIND EROSION

(to the tune of "Blow Ye Winds in the Morning")

In regions that are dry and where the surface of the ground
Is lacking vegetation but loose particles of rock are found, and
Loose material blowing can be much like sandblasting
And wind erosion can make rock formations that are interesting

> Chorus:
> Blowing wind wears rock away, blowing wind hi ho
> Abrasion and scouring from the wind erodes

In the desert where all loose material is blown away
A desert pavement's all that's left from a process called deflation
Boulders may get polished faces, but they're seldom round
The windward side is flattened, meaning sharper angles will be found.
Chorus

Wind erosion varies with how long and hard it blows
And the size of grain, whether sand or silt or loess, but
When wind is slowed down because of plants or topography
Then sand dunes can be found there,
    but the silt and dust in clouds keep blowing
Chorus

SOIL
(to the tune of "My Grandfather's Clock")

A product of the weathering and erosion is the soil,
    that is oh, so important to man
For thousands of years the results of his toil
    has been what he has grown on the land
Grains and fruits, vegetables, grass for his animals
Soil for agricultural needs
Foundation of civilization
Soil is important indeed

    Chorus:
    There's millions of bacteria, and fungi, and
    Many other living things hard to see and
    Soil is formed by the weathering of rocks
    With organic matter over much time

For soil to be formed other factors to recall,
    include what kind of rock it came from
Called parent material with the kinds of minerals
    that affect what the soil will become
Climate (temperature and rain) and the slope of the terrain
Solid particles of sand, silt, and clay
With pores that are either filled with air
Or with water at any one time
Chorus

A profile and colored layers that it shows:
    are horizons of A, B, and C
Humus from plants decomposed makes the topsoil
of A more fertile than B
The minerals on top leaching down will likely stop
At the C-horizon (partially weathered rock)
Laterites—climate wet; pedalfers—temperate
Pedocals where the climate is dry
Chorus

TOPOGRAPHIC MAPS
(to the tune of "Life on the Ocean Waves")

Topographic maps help to show the shape of the land
With contour lines and symbols you'll be able to understand
That points that all had the same
Elevation then became
A single, unbroken line
Seldom will branch or intertwine
The circles are hills or holes, and the scale will indicate
The contour intervals; in between you can estimate

>Chorus:
>A map, a map, a topographic map
>A map, a map, a topographic map

A steeper slope is known from the lines being close together
A gentler slope is shown by the spacing being wider
Darker contour lines
Separated by four others
These are the index lines
And they often have little numbers
Showing elevation from sea level, above or below
In meters or feet are given, which the scale will always show
Chorus

Brown lines will point upstream when crossing rivers or creeks or valleys
And as confusing as it may seem, the branching of blue tributaries
Will point where water flows
And one more thing to know is...
>The U.S. Geological Survey is a governmental agency
>That surveys, maps, and studies landforms and geologic activities
>And puts benchmarks at certain locations, and

The U.S.G.S. has colors, on their topo maps are seen
Black is for what's man-made; water is blue; and plants are green
Chorus

NOTES

[1] *Earthquakes Hazards Program.* February 26, 2003. U.S. Geological Survey. November 2002. <http://earthquake.usgs.gov/4kids/facts.html>.

[2] *Earthquakes Hazards Program.* February 26, 2003. U.S. Geological Survey. November 2002. <http://earthquake.usgs.gov/4kids/facts.html>

[3] *Pegmatology 101.* Pegmatite International. March 2003. <http://www.pegmatology.com/l>.

[4] *Measuring the Damage.* 2001. The Center for Science Education. February 2003. <http://cse.ssl.berkeley.edu/lessons/indiv/davis/hs/RichterScale.html>.

[5] Simkin, Tom and Lee Seibert. *Volcanoes of the World.* Tuscon: Geoscience Press, 1994. as quoted by Scott K. Rowland in *Types of Volcanoes.* <http://volcano.und.nodak.edu/vwdocs/frequent_questions/grp9/question666.html>

[6] Strickler, Mike. Physical Geology lectures. November 2, 2002. GeoMania November 2002. <http://jerscy.uorcgon.edu/~mstrick/geology/Geo_Lectures/Geo_Lectures_page.html>.

[7] Larson, Edwin and Birkeland, Peter, *Putnam's Geology Fourth Edition*, New York, Oxford Universtiy Press, 1982.

[8] Press, Frank, and Siever, Raymond, *Understanding Earth*, W. H. Freeman and Company, New York, 2001 p.46.

[9] *Facts about Minerals.* 2003. National Mining Association. Mineral Information Institute. January 2003. <http://www.nma.org>.

[10] Barna, David. *Fourth of July Fireworks Depend upon Minerals.* 1990. U.S. Bureau of Mines. <http://minerals.state.nv.us>.

[11] Barna, David. *Fourth of July Fireworks Depend upon Minerals.* 1990. U.S. Bureau of Mines. <http://minerals.state.nv.us>.

[12] Press, Frank, and Siever, Raymond, *Understanding Earth*, W. H. Freeman and Company, New York, 2001 p.188

[13] Press, Frank, and Siever, Raymond, *Understanding Earth*, W. H. Freeman and Company, New York, 2001 p. 254.

[14] Press, Frank, and Siever, Raymond, *Understanding Earth*, W. H. Freeman and Company, New York, 2001 p. 261.

[15] Sandburg, Carl, *The American Songbag*, Harcourt, Brace & World, Inc.: New York, 1927, p. 8.

[16] Press, Frank, and Siever, Raymond, *Understanding Earth*, W. H. Freeman and Company, New York, 2001 p. 331.

[17] Press, Frank, and Siever, Raymond, *Understanding Earth*, W. H. Freeman and Company, New York, 2001 p. 339.

[18] Bates, Robert L., and Julia Jackson, ed. *Glossary of Geology.* Alexandria: American Geological Institute, 1987.

BIBLIOGRAPHY

Bates, Robert L., and Julia Jackson, ed. *Glossary of Geology*. Alexandria: American Geological Institute, 1987.

Bishop, Margaret S., and Phyllis G. Lewis, and Berry Sutherland. *Focus on Earth Science*. Columbus: Charles E. Merrill Publishing Co., 1976.

Bramwell, Martyn. *Book of Planet Earth*. New York: Simon and Schuster, 1991.

Campbell, Ann-Jeanette and Rood, Ronald. *The New York Public Library Incredible Earth, A Book of Anwers for Kids*. New York: Stonesong Press, John Wiley and Sons, Inc., 1996.

Field, Nancy, and Adele Schepige. *Discovering Earthquakes*. Middleton: Dog-Eared Publications, 1995.

Fronk, Robert, and Linda Knight. *Earth Science*. Austin: Hold, Rinehard and Winston, Inc., 1994.

Parker, Steven. *The Earth and How it Works*. London: Dorling Kindersley, Inc., 1989.

Press, Frank, and Siever, Raymond. New York: *Understanding Earth*. W. H. Freeman and Company, 2001.

Sandburg, Carl. *The American Songbag*. New York: Harcourt, Brace&World, Inc., 1927.

Sattler, Helen Roney and Maestro Giluio. *Our Patchwork Planet*. New York: Lothrop, Lee and Shepard Books, 1995.

Strahler, Arthur. *Physical Geography*. New York: John Wiley and Sons, Inc. 1969.

World Wide Web Resources

Barna, David. *Fourth of July Fireworks Depend upon Minerals*. 1990. U.S. Bureau of Mines. <http://minerals.state.nv.us>.

Earthquakes Hazards Program. February 26, 2003. U.S. Geological Survey. November 2002. <http://earthquake.usgs.gov/4kids/facts.html>.

Groundwater Basics. February 2003. The Groundwater Foundation. February 2003. <http://www.groundwater.org/GWBasics/gwbasics.htm>.

The Learning Web. August 2002. U.S. Geological Survey. December 2002. <http://www.usgs.gov/education/>.

Measuring the Damage. 2001. The Center for Science Education. February 2003. <http://cse.ssl.berkeley.edu/lessons/indiv/davis/hs/RichterScale.html>.

Michon, Scott. *Alfred Wegener*. 2003. Strange Science, The Rocky Road to Modern Paleontology and Biology. February 2003. <http://www.strangescience.net/>.

Rowland Scott K., *Types of Volcanoes* from *Volcano World*. January 2002. University of North Dakota. December 2002. *<http://volcano.und.nodak.edu/vwdocs/vwlessons/volcano_types/index.html>*.

Strickler, Mike. Physical Geology lectures. November 2, 2002. GeoMania. November 2002. <http://jersey.uoregon.edu/~mstrick/geology/Geo_Lectures/Geo_Lectures_page.html>.

Volcano Hazards Program. Jan 1, 2002. U.S. Geological Survey. November 2002. <http://volcanoes.usgs.gov/Products/Pglossary/pglossary.html>

INDEX

A
aa 31
ablation 76
abrasion 9, 56
absolute age 10
accumulation 76
active 35
aeration zone 68
aftershocks 27
agriculture 85
alluvial fan 20
alpine glaciers 75
altitude 89
angle of repose 71
anthracite 49
anticlines 20
aquifer 68
aretes 77
artesian well 68
ash 33
asthenosphere 14
atoms 37
avalanches 72

B
banding 51
basalt 31, 43
base level 64
batholiths 23
bedding 47
bed load 60
bedrock 41, 72
benchmarks 90
bituminous coal 49
body wave 26
braided channels 8
braided stream 64
breccia 48
Bretz Floods 10
brown coal 49
buttes 22

C
calcium carbonate 69
caldera 29
calving 76
carbonates 39
carbonic acid 69
carrying capacity 60
cataclysm 9
catastrophe 8
catastrophism 10
caves 69
cementation 47
chemical composition 38

chemical formula 37
chemical sedimentary rocks 49
chemical weathering 55
chert 49
cinder cone 29, 32
cinders 32
cirques 77
clastic sedimentary rocks 48
clasts 48
claystone 48
cleavage 39
coals 49
coarse grain 44
color 38
column 69
compaction 47
composite 33
composite volcano 29
compound 37
compression 19, 21, 22
condensation 59
conduits 32
cone 29
confluence 64, 90
conglomerates 48
consolidated 71
consumers 86
contact metamorphism 52
continental drift 13
continental glaciers 75
continental plates 15
contour interval 90
contour line 89
convergent boundaries 16
core 14
country rock 23
crater 29
creeps 71
crevasses 77
cross-bedding 47
crystal 38

D
Death Valley 20
debris flows 72
decomposers 86
decomposition 55
deflation 83
deformation 10, 19
delta 64
density 39
deposition 60, 61, 63
deposition site 47
desert pavement 83
dikes 23

dip 21
disintegration 55
dissected mountains 22
dissolved load 60
divergent boundaries 16
divergent zones 16
divide 64
dormant 35
drainage basin 64
drift 78
drumlins 78

E
earth flows 72
earthquake 25
element 30
elevation 89
epicenter 25
erosion 60, 63, 81
erratics 9, 78
eskers 78
evaporates 49
evaporation 59
exfoliation 56
extension 19, 21, 22
extinct 35
extrusive igneous rocks 43

F
faults 17, 20–21
fault block mountains 20
fault plane 21
fault zone 17
faulting 19
felsic 45
felsic lava 30, 33
fertile 87
fine grain 44
firn 75
fissure eruptions 32
fjords 77
flint 49
floodplain 64
flows 71, 72
focus 25
folded mountains 19
folding 19
foliated 51
footwall 21
forces 19
foreshocks 27
fossils fuels 49
fracture 20, 39
frost action 56
fumaroles 35

Appendix

G
gemstones 39
gemstones 37
geology 7
geysers 35
glacial budget 76
glacial flour 77
glacial ice 75
glacial speed, 76
glaciers 75
glassy 44
gneiss 51, 52
grabens 21
graded bedding 47
grain 44
granite 43
granular 52
gravel 48
gravel bars 9
Great Rift Valley 16, 29
groundwater 66–69
gypsum rock 49

H
hanging valleys 77
hanging wall 21
hardness 38
headwaters 63
heft 39
horizons 87
horns 77
horsts 21
hot springs 35
hotspot 31
humus 61, 86
hydrologic cycle 59
hydrology 59
hydrolysis 57
hypocenter 25
hypothesis 10

I
ice cap 75
ice dome 76
ice sheets 75
ice shelf 76
icebergs 76
igneous rock 41–45
impermeable 67
index contour lines 90
industrial minerals 40
industrial-type metals 40
infiltration 59, 68
inorganic 38, 85
intermediate 45
intermediate lava 30, 34
intrusion 22, 23

intrusive igneous rocks 43

J
J Harlan Bretz 8
joint 20

K
kames 78
kettle lake 79
Kettles 78

L
laccoliths 23
landslides 71, 72
lateral faults 17
laterites 87
lava 29
lava blocks 33
lava bombs 33
lava composition 30
lava domes 34
lava falls 32
lava fields 32
lava flow 29
lava fountains 32
lava plateaus 32
lava tubes 32
law 10
leaching 87
legend 90
levees 64
lignite 49
lithification 47
lithosphere 14
load 60
loam 87
loess 82
luster 39

M
mafic 45
mafic lava 30, 31
magma 14, 29
magnitude 26
mantle 14
map key 90
marble 52
mass movement 71
mass wasting 71
meanders 64
meltwater 78
Mercali Scale 27
mesas 22
metallic luster 39
metallic minerals 40
metamorphic rock 41, 51
Mid-Atlantic Ridge 16

mid-ocean ridge 16
mineral compounds 30
mineral element 37
mineralogists 37
minerals 36–40
Missoula 10
moraines 78
moraines 75
mouth 63, 64
Mt. St. Helens 34
mudflows 72
mudslides 72
mudstone 48

N
nonfoliated 52
nonmetallic 39
nonporous 67
normal fault 21
nutrients 87

O
oceanic plates 15
opaque gems 39
organic 85
organic limestone 49
organic sedimentary rocks 49
organisms 85
Original horizontality 10
outcrop 41
outwash 78
overturned fold 20
oxbow lakes 64
oxidation 57
oxides 39

P
p waves 26
pahoehoe 31
Pangaea 13
parent material 85
parent rock 51
peat 49
pedalfers 87
pedocals 87
pedology, 85
pegmatites 23
permafrost 79
permeable 67
phosphates 39
phyllite 52
physical weathering 55
pillow lava 31
plastic flow 76
plate boundaries 16
plate tectonics 13
plateau 22

plates 14
plunge pool 8
plutonic rocks 43
polar cap 75
pores 59, 86
porosity 67
porous 61, 67
potholes 8
precious metals 37
precipitates 49
precipitation 59, 61
pressures 19
primary waves 26
producers 86
profile 87
pyroclastic flow 33
pyroclastics 29

Q
quartz 37
quartzite 52

R
rapids 63
recrystallize 51
regional metamorphism 51
relative age 10
reservoirs 59
retreating 76
reverse fault 21
rhyolite 33
Richter Scale 27
ridges 16
rift valleys 16
ring of fire 17
ripples 47
river basin 64
rock cycle 41
rock flour 77
rock-formers 37
rockfalls 71, 72
rockslides 71, 72
root pry 56
runoff 59, 60

S
s waves 26
saltation 83
San Andreas Fault 17
sand dunes 83
sandbars 64
sandstone 48
saturated 60, 67
saturation zone 68
scale 90
schist 52
scientific method 7

scoria 32
sea floor spreading 16
secondary waves 26
sediment 10, 47, 60
sedimentary rock 46–49
sedimentation 47
seismic sea waves 27
seismic waves 25
seismograms 26
seismograph 26
seismologists 26
serpentine 52
shale 48
shear 22
shear force 19, 21
shear waves 26
shield volcano 29, 31
silicates 30, 39, 45
silicon 30
sills 23
siltstone 48
sinkholes 71, 73
slate 51, 52
slip 21
snow 75
soil 84–87
solution 60
source 63
specific gravity 39
speleology 69
spreading zones 16
spring 68
stalactites 69
stalagmites 69
stocks 23
strata 47
stratification 47
stratified 61
stratovolcano 33
streak 39
striations 77
strike 21
strike-slip faults 17, 21
subduction zone 16
sublimation 76
subsidence 71, 73
sulfates 39
sulfides 39
superposition 10
surface waves 26
suspended load 60
symbols 37
synclines 20

T
talus 72
tectonic plates 15

tension 19, 21, 22
tephra 33
terminus 77
texture 44
theories 7, 10
till 78
topo maps 89
topographic maps 89
topography 56, 61, 85
transform boundaries 16
transparent gems 39
transportation 63
trenches 16
tributaries 90
tributary 64
tributary glaciers 77
tsunamis 27
tuff 33
turbulence 81

U
u-shaped valleys 77
unconsolidated 71
uniformitarianism 8, 10
U.S. Geological Survey 90
unsaturated 68
unsaturated zone 68
unstratified 78
uplift 19
USGS 90

V
valley 75
velocity 60
vent 29
viscosity 31
viscous 30
volcanic blocks 33
volcanic bombs 33
volcanic dome 34
volcanic glass 44
volcanic islands 31
volcanic necks 23
volcano 28–36
volume 60

W
water table 68
water cycle 59
waterfall 8, 63
watershed. 64
waterways 62–65
weathering 55, 60